MICROWAVE

RADIO

TREVOR MANNING

Technical illustrations by Sylvester Singarayer
Caricatures and drawings by Valkyriasart
Designed and Typeset by JOSEP Book Designs
joseworkwork@gmail.com
Hardcover ISBN 978-0-6481915-6-8
Paperback ISBN 978-0-6481915-5-1
Ebook ISBN 978-0-6481915-4-4

CONTENTS

INTRODUCTION

Microwave radio has been around for over 70 years, so why is there a need to write a new book about it? One reason is that digital radio is so widely misunderstood. For example, despite us calling the topic *digital microwave*, the RF (Radio Frequency) signal is still analogue, yet analogue technology is often no longer taught in our universities and colleges. With the proliferation of the internet as a source of learning, many myths and inaccuracies have become mainstream. Even the name has its controversies. Words such as RF, radio, microwave, millimetre waves and wireless all get used interchangeably with no clear distinction about their differences. It should be noted that in this book the term *microwave radio* will be used as a generic term to cover equipment from 400 MHz all the way up to 450 GHz, for reasons that will be explained later. At the back of the book, there is a list of acronyms to refer to if any abbreviations used in the book are unfamiliar to you.

Another reason to write about microwave is that it is about to get a new boost in life. Back in 1991 when I was studying to become a microwave expert, I recall a friend of mine saying that I was crazy to invest time in a technology that was dying. *'In ten years' time fibre optics will be everywhere, and you will be an expert of an obsolete technology,'* he predicted. Here we are in 2019 and not only is microwave *not* obsolete, but it is growing. For example, in a report published in Dec 2018, Ericsson predicted that 65 per cent of radio sites (excluding North East Asia) would be connected with microwave by 2022. With the expected exponential growth in millimetre wave equipment and with new frequency bands being planned all the way up to 450 GHz that will offer fibre-like capacities, the growth in microwave is unlikely to slow down anytime soon.

Microwave radio planners usually rely on *software tools* to design and plan links, so very little expertise is required to get a link working. The downside of this convenience is that very few radio planners understand the formulas programmed in the software, or even understand the true meaning of the input variables that they use, and so can apply very little judgement when applied to challenging scenarios. Further, international standards have not kept up with equipment advances, and so outdated formulas are coded into radio planning software. Many of these equations are also based on empirical formulas that are only relevant within a defined range. With the new applications for microwave in higher frequency bands, as well as the trend to stretch radio links as far as possible, the formulas are not necessarily accurate. Planning radio links with a blind confidence in the software can result in unnecessary expenditure as well as a degradation in network performance.

Digital microwave systems are incredibly robust. They can run error free under even the most ardent fading situations. Paradoxically, this has become a disadvantage. In analogue networks, if a radio was poorly designed, it was evident on the day of commissioning. In digital radio installations, the design flaws are hidden and only show up when tested in poor weather conditions. Considering this often occurs months after the network is handed over operationally, the result can be a poorly performing network. Microwave, as a technology, is blamed, where it is, in fact, poor design and installation.

Many myths about microwave have to be unlearnt if you have any hope of understanding how to design a robust network, as well as utilising the significant advancements in microwave radio equipment available on the market today.

This book attempts to provide a summary of the basics of microwave radio in a simplified, yet accurate manner. This is no easy task! I recall back in my university days that my physics professor told a story about our understanding of electricity. If you ask a small child how an electric light bulb works they will answer. 'That's easy. You just switch the light on. Everybody knows that'. Ask a schoolchild, and they will answer that it is a little more complicated than that. They will talk about the wire that connects the switch to the light and explain about electron flows in the conductor. They will also wax eloquent about the resistance of the filament and heat emissions. Ask a university professor, and they will answer that no-one really knows. 'We think it has something to do with wave theory, but we also think it might actually be particles, something we call wave-particle duality'. They will then go on about Heisenberg's uncertainty principle and try to explain quantum mechanics to you. At some point getting an accurate answer, is less useful than one that is slightly inaccurate but at least helps you in a practical way. In this book, I will only dispel the industry myths and inaccuracies where it really matters. This is a practical book. It is not an academic treatise.

I wrote this book to be a handy authoritative reference guide if you are involved in planning, designing or operating a microwave radio network. It is also intended to provide a novice with a reliable and simple-to-understand overview of microwave radio systems written by someone in the industry who is experienced from both an operator and a supplier perspective.

MICROWAVE FUNDAMENTALS

- ❖ Understand the history of microwave
- ❖ Explain the fundamentals of electromagnetic waves
- ❖ Know the true risks associated with microwave fields and identify the safety precautions needed for microwave technicians
- ❖ Understand the basics of spectrum planning
- ❖ Describe the benefits and disadvantages of microwave radio links, compared to other transmission mediums

HISTORY OF MICROWAVE

In 1873, Scottish professor James Maxwell published, '*A Treatise on Electricity and Magnetism*', where he mathematically predicted the existence of electromagnetic waves. Before this, magnetism and electricity were believed to be two different fields of science. It was considered a strange coincidence that the formulas for magnetism and electricity were linked by a constant factor, which was 'close to the speed of light'. Maxwell proved that the constant factor was, in fact, the speed of light, because electric and magnetic fields mutually induce energy into a common wavefront called an electromagnetic signal, which travels at the speed of light. Around the same time, Alexander Graham Bell was doing his experiments with telephony. These two discoveries could be regarded as the birth of wireless telecoms.

In 1886, Heinrich Hertz did a series of practical experiments to prove Maxwell's theory and inadvertently invented the world's first antenna. What he did was to set up an oscillator at one end of his lab and injected a known frequency into a wire with a spark gap. Hertz then detected the signal through a similar device at the opposite end of the lab. As he knew the frequency of oscillation, by measuring the wavelength, he could deduce the speed of light. (Wavelength is inversely proportional to frequency by a constant factor c, which is the speed of light). Hertz's contraption is the basis of a modern dipole antenna, which is also used as the driver element for a household rooftop TV *yagi* antenna. Guglielmo Marconi was the first of these men to see the practical and commercial significance of these inventions, and he began producing equipment through his Wireless Telegraph and Signal Company, which is now part of Ericsson.

The first long-haul microwave link was built between New York and Boston in 1947. Commercial point-to-point microwave radios only became mainstream in the 1950s, where they were used for the long-haul, high capacity networks used to transport carrier-grade voice and data circuits. By the end of the century, many of the long-haul links had been replaced by fibre optic networks which offered much higher capacities. The explosive growth in microwave started in the 1990s in Europe, where new cellular GSM networks used microwave radio to build out their networks. Microwave radio growth was also fuelled by so-called last mile access, where it acted as a fibre-in-the air to connect customers' premises to the network. It was also the transmission medium of choice for mobile carriers doing fast network rollouts, in greenfield areas, where fibre was costly and time-consuming to establish. Today microwave radio capacities are available that exceed 10 Gbps. At these capacities, microwave radio is usually cheaper than fibre unless the fibre is already installed, and microwave is much faster to deploy as it bypasses the time-consuming and costly servitudes (wayleaves) required to lay cable runs. Microwave radio is an ideal technology to combine with fibre optics as a redundancy solution for ultra-reliable networks as well as using it to exploit its ultra-low latency (delay) characteristics. For this reason, it is likely to continue to grow in support of the expected explosion in new networks such as 5G and Internet of Things (IoT) networks.

FUNDAMENTALS OF ELECTROMAGNETIC (EM) WAVES

Many textbooks draw a microwave signal as a *ray*, leading to an incorrect assumption that the microwave signal travels like a laser beam over the air. Understanding how a microwave signal travels is fundamental to understanding why microwaves usually need so-called line-of-sight to work reliably and is also fundamental to understanding why interference signals cannot be ignored even when there is no line-of-sight.

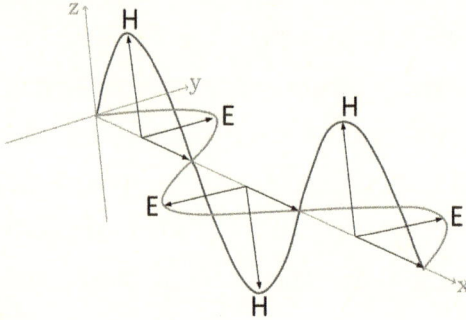

Figure 1.1 Electromagnetic wavefront

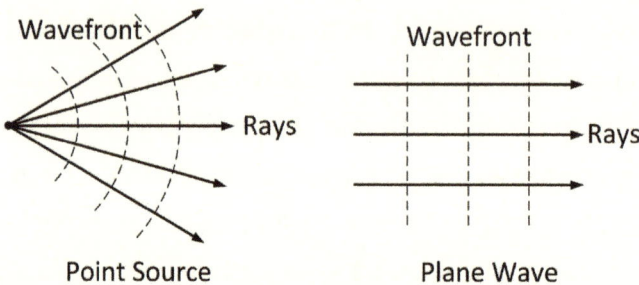

Figure 1.2 Wavefront and ray representation

A microwave radio signal travels as a transverse electromagnetic (TEM) wave. Light is also an electromagnetic wave in the visible light spectrum. Many of the formulas and theory of microwave propagation originate from the physics of light waves. The periodic sinusoidal wavefronts of the orthogonal magnetic field strength (H) vectors and electric field strength (E) vectors are uniform and both travel transversely to the direction of propagation.

The sinusoidal representation, as well as a simplified ray model, is shown in Figures 1.1 and 1.2.

The ray and wavefront models are provided as a simplification to the complex TEM wavefront, but it should always be remembered that in practice, the signal is not contained within these lines.

The basic characteristics of the microwave signal are defined by analysing the sinusoidal wavefront, as shown in Figure 1.3.

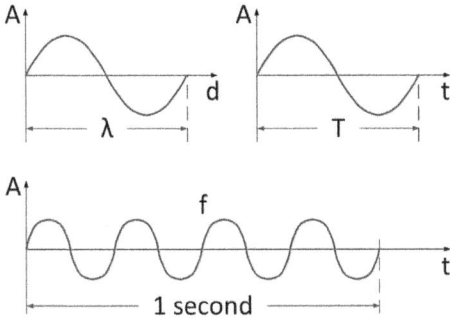

Figure 1.3 Sinusoidal microwave signal

The **period (T)** is the time it takes for the signal to complete one cycle of the microwave sinusoidal signal. Period is usually measured in seconds per cycle.

The **frequency (f)** is the inverse of the period. In other words, it is the number of cycles in a one-second period.

$$f = 1/T$$

Frequency is sometimes confused with the speed of the signal. It will be shown later that the speed of the signal is the same irrespective of frequency. If you could physically see a microwave signal and started and stopped a stopwatch in one second, the frequency would be the number of cycles that you counted during that one second period. The higher the number of oscillations, the higher the frequency.

The **wavelength (λ)** is the spatial length of one cycle. Maxwell proved that in a vacuum, this is the speed of light (**c**) divided by the frequency.

$$\lambda = c/f$$

What the formula proves is that wavelength and frequency are inversely proportional. As the frequency increases, the wavelength gets smaller, which has a practical significance in microwave links. For example, the gain of a parabolic antenna increases with frequency because more wavelengths fit into the same physical reflector space.

The **velocity of propagation (v)** is a function of the medium in which the signal travels. In a vacuum, the microwave signal travels at the speed of light. In any other medium, the signal slows down proportional to the dielectric constant (relative permittivity) (ε) of the medium it is travelling in.

$$\text{RF delay} = \text{Distance} / (\text{VF} * c)$$

where Velocity Factor (VF) $= 1/\sqrt{\varepsilon}$

It will be seen later that the practical significance of velocity factor (VF) is that microwave rays travelling over the air travel at different speeds at the top of the beam to those at the bottom of the beam which causes bending of the wavefront. Increased latency is also introduced when using a cable containing anything other than air due to the density of the medium. For example, a microwave signal transported in a foam-filled Heliax cable could travel 20 per cent slower than the same signal in an air-filled waveguide.

One benefit of a microwave signal travelling over the air with a waveguide or slip-fit connection to the antenna is that the signal travels close to 100 per cent of the speed of light. A fibre optics signal only travels at approximately 70 per cent of the speed of light in practice due to internal reflections and refraction through the glass fibre, making microwave signals the lowest latency transmission medium on the planet. The low latency of microwave radios is sometimes taken advantage of by High Frequency Trading (HFT) institutions, where slashing

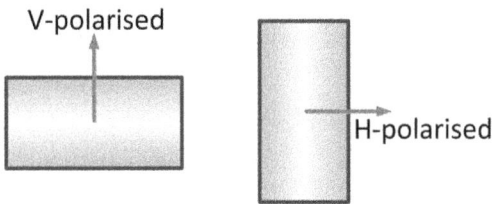

Figure 1.4 Polarisation of horn-feed

milliseconds off a transaction time can translate to millions of dollars in a financial year.

The **polarisation** of the signal is defined by the orientation of the electric field strength (E) vector. In a rectangular waveguide, the electric field is dominant on the longest side. Hence, an antenna with this orientation of horn-feed is vertically polarised. If the antenna is turned 90 degrees, it would be horizontally polarised as shown in Figure 1.4.

From a practical point of view, each polarisation can be treated as a separate carrier frequency. It will be seen later that a configuration called Co-Channel Dual Polarisation (CCDP) allows doubling of the capacity on a single frequency by using each polarisation of the signal. Horizontal polarisation performs worse in a rainstorm and also experiences worse reflections. On any established link the polarisation needs to match. In other words, the antennas need to be accurately aligned as horizontal to horizontal or vertical to vertical. Cross-polar discrimination can be exploited to reduce interference, which will be considered in Chapter 6 (Frequency Planning).

SAFETY OF MICROWAVE SIGNALS

Microwave radio waves are part of the electromagnetic spectrum as shown in Figure 1.5. Microwave signals are below the visible spectrum and importantly, are also

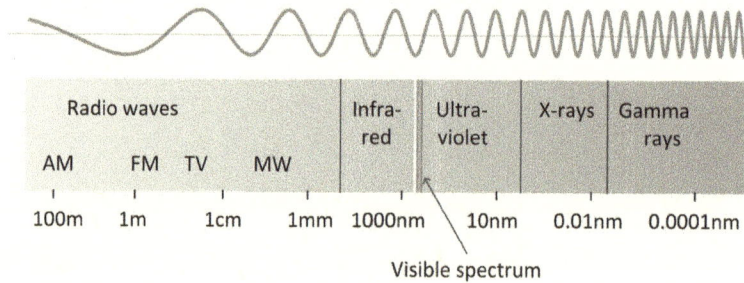

Radio waves				Infra-red	Ultra-violet	X-rays	Gamma rays
AM	FM TV		MW				

100m   1m   1cm   1mm   1000nm   10nm   0.01nm   0.0001nm

Visible spectrum

Figure 1.5 Electromagnetic spectrum

below the ionised spectrum where there is harmful radiation, such as X-rays and gamma rays.

Microwave frequencies are thus non-ionised and therefore lack the energy to change the DNA composition of human tissue and thus are unable to cause cancer. A lot of research has gone into the potential dangers of the heating effect from EM radiation and whether this could accelerate the growth of cancer cells. So far, there is no proof to suggest that this is possible even with mobile phones that transmit very close to the human head, or from TV transmitters that transmit kilowatts or even megawatts of energy. In the case of point-to-point microwave, the signal strength is incredibly low - less than a watt at the source - and the signal falls off exponentially with distance, making the signal almost too small to measure even metres away from a transmitting antenna. The allowable, safe leakage through a microwave oven door is far higher than the transmit signal at the antenna of a microwave link. This transmit signal is also located at the antenna source, which is on a rooftop or radio tower. Considering the signal strength was very low in the first place, by the time the significantly attenuated signal has human contact, the levels of microwave exposure are minuscule.

Useful advice and guidance on the health and environmental effects of non-ionising radiation (NIR) are provided by ICNIRP (International Commission on Non-Ionising Radiation Protection). Compliance to their guidelines ensures the protection of people and the environment from the detrimental NIR exposure. While the risks to the general public from microlink exposure are minisculely small, for reasons discussed previously, there are some practical considerations for people working on microwave equipment within telecoms. For example, for riggers who climb TV towers to install or repair microwave equipment, protective suits should be worn when climbing through high-risk radiation zones. Alternatively, the TV signal should be turned down during maintenance periods. The reason is that TV or radar signals transmit a far higher power than microwave point-to-point links.

The human eye and the testes are very sensitive to radiation, and so one should never look into an open transmitting waveguide or stand in front of a transmitting test antenna for long periods.

MICROWAVE SPECTRUM TERMINOLOGY

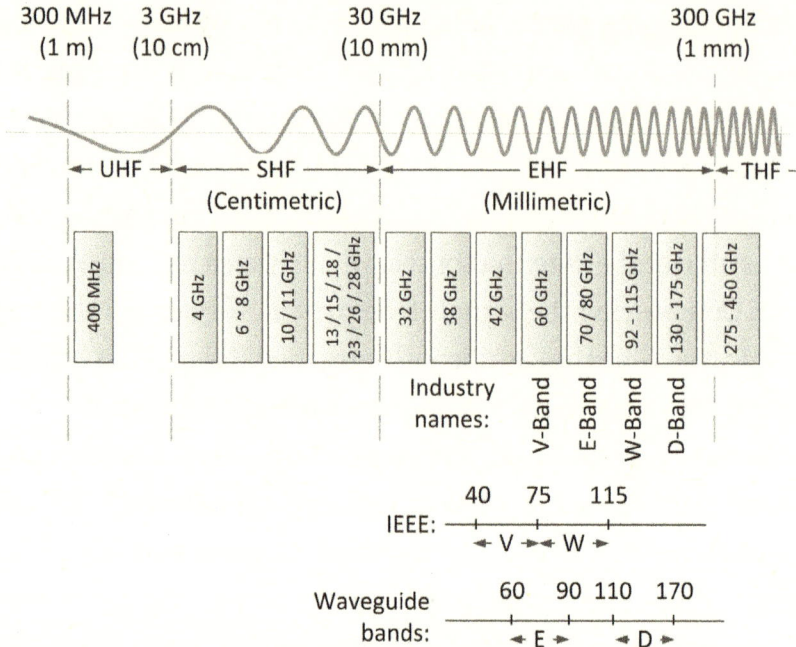

Figure 1.6 Naming of microwave frequency bands

When we talk about microwave spectrum, we are referring to the band of frequencies that are used for microwave transmission signals. The industry uses confusing naming conventions derived from a mixture of the ITU, IEEE, radar and waveguide bands. For example, people talk about C or X-band (from IEEE standards) when describing satellites, E-band (from waveguide standards) when describing microwave links in 70/80 GHz frequencies, V and W (from IEEE standards) for the new high-frequency millimetre wave bands and yet still refer to UHF and SHF (from ITU standards), for radio bands.

Some factual information is provided next to help decipher the mess. Refer to Figure 1.6 to assist with understanding the naming conventions discussed in the following section. For those readers not interested in the gory details of spectrum bands and naming conventions, feel free to skip to the next section called Microwave Spectrum Fundamentals.

The ITU refers to 3-30 GHz as SHF (Super High Frequency) and calls the signals **centimetric waves** because signals from 3 GHz to 30 GHz have wavelengths between 1 cm to 10 cm.

Frequencies from 30 GHz to 300 GHz are referred to as EHF (Extremely high frequency) and are called **millimetric waves** because the wavelengths vary between 1 mm to 10 mm.

Frequencies in the UHF band are from 300 MHz to 3 GHz and are called decimetric waves. However, this terminology is seldom used in the industry. 400 MHz is the lowest frequency band allocated to microwave applications, and the ITU has recently started considering developing microwave channel plans from 275 GHz to 450 GHz, which falls within the so-called Tremendously High Frequency (THF) band.

In practice, the definitions are not so precise. Various literature refers to millimetric wave bands as mmWave or millimetre wave bands. Also, the GSMA refers to 26 GHz, 40 GHz, and 66-71 GHz all as millimetric bands, while some industry literature only refers to millimetric bands when the frequencies are above 55 GHz, thus excluding 38 GHz. The reason is that traditional microwave bands are from 4 GHz to 42 GHz, and so 38 GHz is generally classified in this category.

The IEEE V-band extends from 40 - 75 GHz, but the microwave industry is usually referring only to the unlicenced 60 GHz band (57 - 71 GHz) as V-band.

The IEEE W-band is from 75 GHz to 115 GHz, but the microwave industry generally refers to W-band from 92 GHz - 115 GHz because the 70/80 GHz is now universally referred to as E-band. Technically speaking, the E-band covers 60-90 GHz and is not an ITU or IEEE microwave band as it derives its name from waveguide terminology. The waveguide band from 110 GHz to 170 GHz is called D-band. The new microwave millimetric wave band from 130-175 GHz, which is yet to be channelised by ITU, is often, therefore, referred to as D-band.

Another misunderstanding is related to the spectrum that is designated for microwave backhaul versus spectrum for 5G access. Traditional long-haul microwave bands include 4 GHz, 6 GHz, 7 GHz, 8 GHz, 10.5 GHz, and 11 GHz with medium to short-haul bands including 13 GHz, 15 GHz, 18 GHz, 23 GHz, and 38 GHz. In the last ten years 26 GHz, 28 GHz, 32 GHz, 42 GHz, 60 GHz, and 70/80 GHz, and others, have been added.

In 5G, planners have an insatiable need for more spectrum, and so all frequency bands will likely be targeted. Of immediate interest to 5G is 600 MHz, 700 MHz, 2.3 GHz, 2.6 GHz, 3.5 GHz, 4 GHz, 6 GHz, 24 GHz, 26 GHz, 28 GHz, 37-40 GHz, 42 GHz, 48 GHz, and 60 GHz. These bands are usually auctioned to specific operators for millions of dollars and so are not available for use in microwave backhauling. There is strong support to keep E-band (70/80 GHz), 38 GHz and 32 GHz for microwave backhaul use, which also provides alternative spectrum for microwave links currently deployed in bands earmarked for 5G access. With the loss of bands such as 26 GHz to 5G access, there is an interest for microwave backhaul in the much higher frequency spectrum, such as W-band (92 GHz to 115 GHz) and D-band (130-175 GHz), all the way up to 450 GHz.

It should be noted that there has been a trend to converge fixed and mobile access, and spectrum is being repurposed to meet the needs of the industry, with some countries, such as

US, Japan, South Korea defining non-ITU bands to meet their needs.

The last thing to clear up is the various terms that are used for radio systems. **Wireless** is an industry term used to differentiate between cable or satellite networks and includes lower frequency cellular radio bands and even Wi-fi (Wireless Fidelity). Frequencies associated with Land Mobile Radio (LMR), such as 400MHz or 150 MHz, are often referred to as **radio frequency bands**, to differentiate them from microwave bands. In the physics books, anything above audio signals (20 kHz) are considered **Radio Frequency (RF)**.

In this book, the term **microwave radio** will be used as a generic term to cover all fixed point-to-point equipment from UHF frequencies (400 MHz) to above millimetre wave bands (up to 450 GHz).

MICROWAVE SPECTRUM FUNDAMENTALS

Microwave links operate in full duplex mode, which means that a separate communication channel is available in each direction allowing simultaneous transmission of data in each direction. Most microwave systems achieve full duplexing by using different carrier frequencies in each direction and by aggregating the user data with Frequency Division Duplexing (FDD). In other words, the traffic channels are stacked next to each other in the frequency domain.

The RF bandwidth is symmetrical irrespective of the asymmetry in data traffic because no distinction can be made between dummy data versus real data. In analogue networks, when no data was present, there was no modulation of the carrier, and thus RF bandwidth was proportional to baseband bandwidth. In digital radios, the full RF bandwidth is used irrespective of baseband bandwidth. This process of modulating traffic, whether it is real or not, can lead to bandwidth inefficiencies if the traffic is asymmetrical. For example, in cellular networks, the downlink traffic often far exceeds the uplink traffic. Manufacturers can produce radios with asymmetric RF channels, but at the time of writing, no Regulators had approved this configuration.

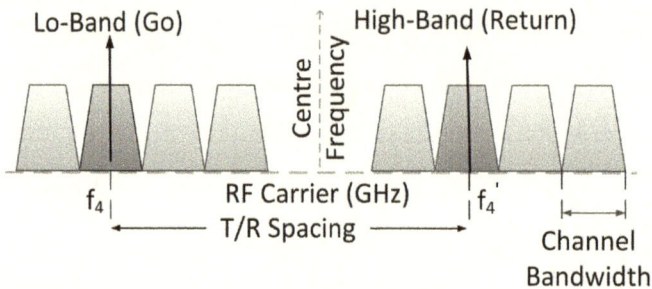

Figure 1.7 A duplex channel plan

The way that the spectrum is organised is that each RF channel is paired with a fixed Transmit to Receive (T/R) spacing, as shown in Figure 1.7. The user traffic that has been modulated onto the carrier signal with an RF frequency of f4 is paired with user traffic modulated onto the return frequency f4' travelling in the opposite direction. By using this organised method of transmission, many simultaneous communication channels can be established without interference.

The transmit signal and receive signal are both connected to a common antenna using a device called a duplexer. This device filters unwanted signals to achieve the required level of isolation of the receiver from its own transmitter, as well as coupling both transmit and receive signals to the common antenna. The layout is shown in Figure 1.8. It can be seen that the transmit interference signal I_{TX} into receiver f4' is reduced by two things. Firstly, the transmit filter, which is tuned to f4 and so attenuates f4' by the value of the filter rolloff, and secondly via the duplexer isolation.

The separation of the T-R frequencies is mainly dependent on the physical ability of the duplexer to stop the transmit signal leaking through into the local receiver, through the common antenna. The larger the T-R spacing, the easier it is for a duplexer to filter out the unwanted leakage signal. It is of

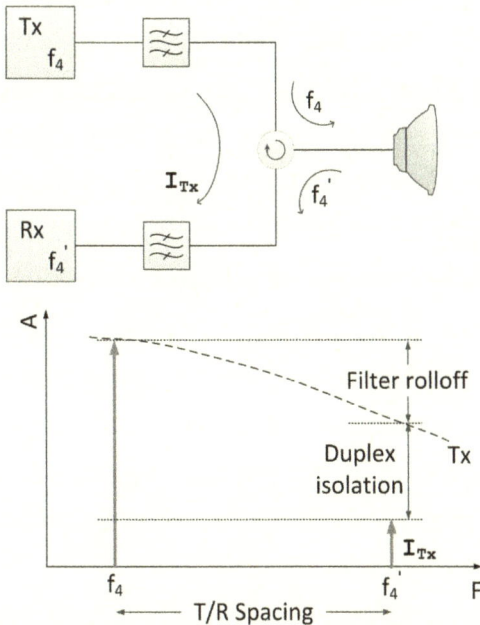

Figure 1.8 Frequency Division Duplexer

interest to note that at higher frequencies such as 23 GHz, the separation required exceeds 1 GHz. Inexperienced planners often forget that there are separate carrier frequencies operating in each direction.

More recently, in the so-called millimetre wave bands, there is increased use of Time Division Duplexing (TDD) for microwave links. In these systems, traffic channels are differentiated in the *time*, rather than the *frequency* domain.

ALTERNATIVE TRANSMISSION OPTIONS

Microwave radio is not the only available transmission medium.

Twisted pair copper cables have been around for decades. With Digital Subscriber Line (DSL) solutions, data can be carried over the same cables used for voice. Copper cable has the benefit of being extensively available, especially in first world countries, but usable capacity is limited by the distance from the local exchange and the quality of the copper. DSL standards are available in both symmetrical (SDSL) and asymmetrical (ADSL) versions. With new technology standards such as VDSL vectoring (Very high bit rate DSL) and G.fast, speeds of up to 1 Gbps are now possible, with speeds up to 10 Gbps already achieved in laboratories using new emerging standards, albeit over short distances (few hundreds of metres). The quality of the copper and distance from the exchange, are key factors to practically achieving these high-speed connections.

Coaxial cable networks, using DOCSIS (Data Over Cable Service Interface Specification) standard, were historically rolled out for TV distribution, due to the superior bandwidth and quality achieved, compared to twisted pair copper cable, and some networks combined this with fibre in so-called HFC (Hybrid Fibre Coaxial) networks.

Fibre optic cables are seen as the utopia of telecoms, due to their almost unlimited bandwidth. Where the cables are not already in place, a new cable can be very costly and time-consuming to build. In some geographic areas, such as mountainous areas and over lakes, microwave radio may be much more practical. In city areas, a 50 m distance between two buildings could end up being hundreds if not thousands of metres of fibre optic cable due to the fibre trench arrangements.

Free Space Optics (FSO) solutions use an infrared laser beam to transmit high capacities over short distances. No licence is necessary, making them ideal for fast rollouts. Optical line of sight is needed, so they can be unreliable where fog is prevalent. Dirt on the lens as well as scintillation from sunlight also needs to be taken into account, as well as pole stability due to the highly directional beam. Even building sway in the wind or expansion and contraction from temperature changes can impact the performance, and so equipment countermeasures are sometimes deployed to overcome these effects.

Satellite systems are widely available, and increased demand has dramatically reduced costs. Despite this, it is still a capacity-limited, and costly solution, where other options are available. Typically, telecoms satellites are geostationary, where they are located at 36,000 km to rotate at the same speed as the earth, so no tracking is necessary. The round trip time delay is called latency, and this is high, due to the transmission distances involved. The signal to the satellite travels at the speed of light. Even at this speed, it takes a few hundred milliseconds to travel such a considerable distance. For critical real-time services, the target for latency figures is below 10 ms, creating quality issues for these services over a geostationary satellite.

Medium and low earth orbit (LEO/MEO) satellites are also used, which have less latency, but do require tracking.

Fixed wireless solutions include **point-to-multipoint (PMP)**, in apparatus-licenced as well as class-licenced microwave bands. The distinction between an individual (*apparatus*) licence and a general licence-exempt (*class* or *unlicenced*) licence will be discussed later in Chapter 6 (Frequency Planning). Point to multipoint is fast to deploy, even for licenced PMP as it only requires a licence at the node. Licensed operation is attractive if an operator can reserve a block of frequencies, but the relatively scarce availability of blocks of suitable spectrum does also limit PMP use. Cost savings can also be achieved through the nodal antennas, as well as the fact that at the node only one device is required, rather than a dedicated device paired to each remote end. Considering the traffic is shared between the various customers at the node, the multiple access solutions can introduce increased latency and reduced capacity, which can impact real-time services.

BENEFITS AND DISADVANTAGES OF MICROWAVE

Point-to-point (PTP) microwave radio is relatively fast to deploy compared to other solutions and has fibre-like performance with ultra-low latency. With the advances in multi-level modulation schemes, baseband compression schemes, dual polarisation operation with cross-polar interference canceller (XPIC), as well as wider RF channels - such as 112 MHz - gigabit per second capacities per channel are possible. Using multiple channels capacities exceeding 10 Gbps are possible.

Another major benefit of microwave, especially in urban areas, is that it offers flexibility. Microwave radios can be repointed or redeployed, as changes in the network occur, significantly decreasing the Total Cost of Ownership (TCO). The redeployed equipment provides improved return on investment (ROI) on the capital expenditure (Capex).

The challenges with microwave are the availability and cost of suitable spectrum, achieving sufficient line-of-sight over obstructions, such as trees, and the limited capacity compared to fibre. Microwave is also an ideal redundancy option. A true redundancy option should have no common point of failure, which is difficult to achieve for dual fibre solutions, especially in non-urban areas. Microwave radio provides true media and route redundancy to fibre, which is essential for critical applications.

CHAPTER IN A NUTSHELL

In summary, what do I need to know about microwave?

Microwave radio is a growing technology and is a crucial part of building new telecoms networks, including 5G. In practice, microwave bands start at 400 MHz and will eventually go up as high as 450 GHz, which will facilitate transmission capacities of tens or even thousands of Gigabits per second.

Historically, microwave radio started in the 1950s and has always been an excellent alternative to fibre, where capacities do not exceed tens of Gigabits per second, as they are often faster to deploy and cost less than building new fibre systems.

A microwave signal travels as a continuous, sinusoidally-varying Transverse Electromagnetic (TEM) wavefront and therefore follows the laws of optical physics from a transmission perspective. The wavelength of the signal is the spatial distance the signal travels before it repeats itself and is inversely proportional to the frequency. The signal travels at the speed of light over the air, irrespective of frequency. Information can be coded onto either, or both, polarisations of the RF signal.

Microwave signals are lower in frequency than light signals, and because they are non-ionised cannot cause cancer. While there are studies to understand if the heating effect of microwave could accelerate a cancerous growth, there is no evidence to suggest it is possible. The probabilities are extremely low as the signal has less power than a light bulb and is far below the allowable leakage through a microwave oven door. The microwave exposure level that would be experienced by anyone in the path of a microwave radio signal is minuscule.

Microwave field technicians should take precautions when working on a radio tower, in close proximity of any high power equipment such as TV or radar and also avoid looking into an open waveguide or having continuous exposure to a transmitting antenna at very short distances in a lab environment.

Microwave Spectrum is the term used to describe the band of frequencies used in microwave. Historically microwave frequency bands were up to 42 GHz. Bands such as V-band (60 GHz) and E-band (70/80 GHz), as well as future bands above 90 GHz (W-band, D-band), offer very high capacities to meet the increasing capacity demands for microwave systems.

Microwave radio systems offer higher capacities over longer distances than traditional copper systems, and do not have the latency concerns of satellite. Where fibre is not already available microwave can provide a similar quality alternative at a lower price point.

Microwave is also an ideal redundancy option to fibre as it provides true media and route redundancy for critical applications.

2

LINK PLANNING

❖ Know how to identify and plan for site acquisition risks
❖ Describe the requirements for an active and passive radio repeater site
❖ Understand the application and risks of using digital data
❖ Network topologies and spectrum considerations
❖ Learn why GIS basics matter
❖ Know how to confirm site coordinates and create a path profile
❖ Know how to establish line-of-sight in the field
❖ Understand and describe the details of site and path surveys

SITE AND PATH PLANNING

A customer decides that they want to establish a microwave radio link so that they can create a Wide Area Network (WAN) Ethernet connection. As a link designer, where do you start?

The first step is to find out exactly where the customer connection is required. Identifying the site location may sound pretty obvious, but the devil is in the detail. Providing a site location is more tricky than may be initially realised. Let's assume it is a factory location. There will likely be multiple buildings on the site and, even once a specific building is located, the specific location of the IT room and the positioning on the roof may have tens to hundreds of metres of uncertainty. In many cases, the end location of the service is not physically built yet at the time the planning and design are required.

In order to start the link planning process, the site location should be mapped relative to the nearest telecom infrastructure. Physical maps can be used, although in most countries digital maps can be used. A Digital Elevation Model (DEM) is a digital map where a raster is superimposed on the earth, creating a grid system with three coordinates (x,y,z). The x and y points represent the physical location of the corner of the pixel within the raster, and the z coordinate represents the height of that pixel block.

A Digital Terrain Model (DTM) is similar to a DEM but usually has terrain clutter added to the z component. Digital Surface Model (DSM) data is also being used increasingly in microwave planning as it has accurate terrain clutter data. For example, LiDAR (Light Detection and Ranging) is a DSM data collection method using reflected laser signals mounted to an aircraft that flies overhead to create 3D geospatial measurements. Accurate clutter data is thus available, with centimetres of accuracy - at least at the time that the data was collected!

The simplest method is to use terrain data from the Shuttle Radar Topography Mission (SRTM). The SRTM data is available free-of-charge from the internet. It should be noted that this Digital Elevation Model (DEM) data has a default of 3 arc-second accuracy (approximately 90 m). In the United States, it is 1 arc-second (30 m). More accurate data is available at a cost in many countries. However, it should also be remembered that in other countries, no suitable DTM or DEM data is available, and topographical maps are still required for line-of-sight predictions.

Digital data should be used with caution when planning a microwave system.

Imagine that you were out hiking, and you staked out an area 90 by 90 m and tried to estimate the average height of the terrain within that square. The job would be reasonably straightforward if the terrain was flat or did not have any abrupt changes in height. On the other hand, if the terrain was undulating and included large rock outcrops and steep cliffs, it would be almost impossible without a computer.

A computer algorithm will easily complete the exercise. The frightening thing is that many designers believe that the single DTM data point downloaded onto the PC planning software for that pixel block - a 90 m by 90 m area in many cases - is the exact terrain height that the microwave signal encounters as it passes that terrain area. In reality, the centre point of the radio path could bypass any portion of the pixel square, and at any angle, yet the DTM value for an entire pixel of ground area is represented by a single point.

The only way to know if the DTM data is a reasonable approximation of the real terrain is to visit the area and conduct a site survey. I have personally done a radio survey at a radio site where a 15 m rock outcrop was not represented by the DTM data, nor was it shown on the topographical map, because the contour spacing tends to be in increments of 20 m. We would have built an inadequate tower had we not visited the site.

In microwave links, moving one end by less than 1 metre may be the difference between a link working or not working.

Doing a site survey from a location that is not at the exact location of where the antenna will end up going, is very risky. Using office-based data from maps or digital terrain data in an urban environment can even result in rooftop surveys being done from the wrong building. Bear in mind that even data to within one second of accuracy still creates an uncertainty of 30 by 30 m in the exact location of the antenna placement. In one multi million dollar European project rollout that I was involved in years ago, more than 75 per cent of the surveys had to be redone due to inaccurate site location issues. In other long-link projects, I have seen site survey people establish line-of-sight to the wrong tower.

Using digital map information is a useful starting point, but even where highly accurate - and expensive - data is available, it is no substitute for doing field surveys where the current condition of both the path and site details can be assessed.

The best way of being sure that you are at the right location and are looking in the right direction is to establish known locations that will be easily identifiable from the site. The exact bearing can then be worked out to verify that you are at the

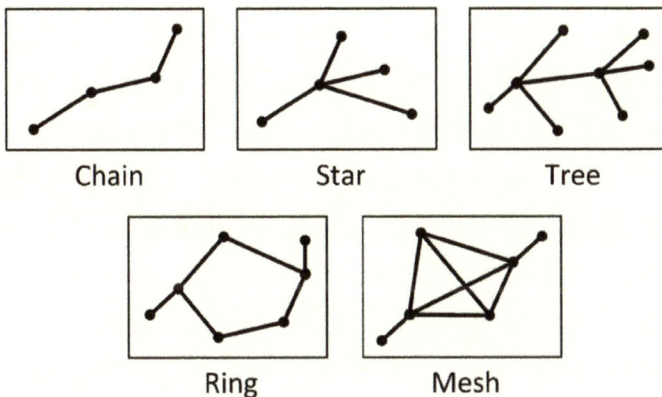

Chain Star Tree

Ring Mesh

Figure 2.1 Network topologies

correct site. If the landmark is at a different bearing when you are at the site, you are probably not in the right place. Knowing the right direction to look in when doing the survey is critically important to avoid the errors mentioned earlier.

At the customer end, the antenna can be mounted on the roof of the site. Alternatively, a tower or pole may be required to achieve line-of-site clearance over local clutter and other obstructions along the path. If direct line-of-sight to a network site is not achievable, an intermediate repeater site may be required.

When planning the network, different geometric topologies may be considered.

Some typical network topologies are shown in Figure 2.1 that include chains, star, tree, ring and mesh layouts. Long microwave chains should be avoided as a transmission break in any of the links affects the whole chain. Hardware protection should be considered when equipment is deployed in chain, tree and star topologies, and no alternative physical path exists for the traffic. In star topologies, the angles between sites need to be carefully considered to mitigate the risk of interference. Tree topologies are a practical mix of chain and star topologies.

It is usually desirable to aggregate hops into a high-availability node, which can be designed with high reliability by having multiple routes into it, and enhanced power system availability. Traffic aggregation and bandwidth efficiency measures can also be more easily done at nodal sites. Mesh is the ideal redundancy topology as it creates multiple physical path options. From a financial perspective, mesh topology is usually only practical in urban areas where link lengths are short, and capacities are high. Ring and mesh topologies provide much greater resilience and create *multiple* Layer 1 physical paths. These paths can be used by Layer 2, or Layer 3 protection schemes, including MultiProtocol Label Switching (MPLS) or Software Defined Wide Area Network (SD-WAN) networks. If a wireless ring topology is used to achieve redundancy, the paths must be chosen with maximum geographical separation to avoid any weather-related outages affecting both primary and secondary paths of the ring.

Software Defined Network (SDN) is a network architecture where the control and data planes are separated. The control plane can thus be centralised and programmed via standard interfaces, independent of the hardware deployed at the edge of the network. SDN works closely with NFV (Network Function Virtualisation), where various networking components such as firewalls are virtualised in software, rather than being dedicated hardware devices. An extension of the SDN concept is to build wide-area networks with a separated data and control plane, called SD-WAN (Software Defined Wide Area Network). In SD-WAN architecture, the data traffic is distributed across multiple physical paths with the traffic controller being located in the so-called Cloud.

Some radio vendors are evolving away from offering pure Layer 2 Carrier-Ethernet implementations to having Layer 3 functionality built into the microwave equipment, including Internet Protocol MultiProtocol Label Switching (IP-MPLS). Hierarchical Quality of Service (H-QoS) is sometimes also offered to make the service more suitable to shared networks with a mixture of voice, data and video traffic. Another consideration in network architecture is the introduction of network slicing, which allows different traffic types to be optimised to run over multiple virtual networks configured for those specific use cases. For example, some applications, such as autonomous cars may require ultra-low latency but have relatively low capacity requirements, whereas another application such as downloading a large file may be delay tolerant but require optimised bandwidth. Network slicing caters for both needs.

HIGH-LEVEL SPECTRUM PLANNING

The details of spectrum planning will be discussed in Chapter 6 (Frequency Planning). At the early planning stages, a number of spectrum considerations need to be taken into account.

The first consideration is which frequency band will be the most appropriate. Table 2.1 shows some typical high-level options, taking into account that the specific frequency choices are Regulator-dependent for each country. The exact

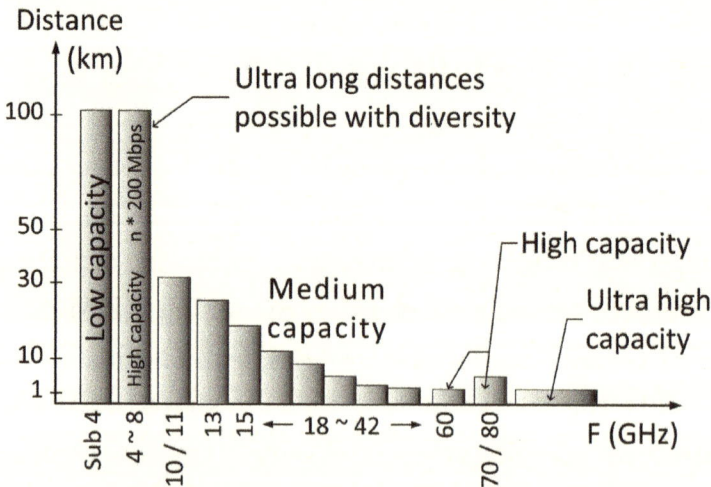

Figure 2.2 Relationship between capacity, link length and frequency band

details will vary for different types of radio equipment that may support different modulation schemes. The capacities suggested also exclude the various baseband optimisation and header compression schemes that maximise bandwidth efficiency. In Chapter 7 (Hardware Considerations) we will be covering concepts such as multi-channel and multi-band options, that will also impact the details.

The purpose of Table 2.1 and Figure 2.2 is to provide a high-level view of how capacity and hop length are related, at the *early* stages of planning. In practice, there will be many exceptions in different locations and different equipment in the final implementation. There is also a global trend to repurpose frequency bands for fixed-mobile converged access. Knowing the cost of spectrum, plus having an understanding of the sorts of capacities available in different bands, is useful during the planning stages. It is also useful to understand any minimum-link-length policies that the Regulator might have, which determines which frequencies may be allowed. Licensed spectrum in high-density areas can be prohibitively expensive in certain bands, and spectrum costs can be the primary driver to use general use (unlicensed) bands. Licensing will be discussed in more detail in Chapter 6 (Frequency Planning).

There is a world-wide tendency for the Regulators to provide more freedom to Operators to make spectrum decisions and even to allow self-coordination in certain bands, such as E-band.

Table 2.1

Microwave frequency band options

Licensed Frequency bands	Planning Considerations
Sub 4 GHz	Historically low capacity (typically <10 Mbps) but being repurposed for fixed-mobile converged access Ultra-long (100 km+) hops possible with low modulation schemes Operation without line-of-sight is possible Spectrum cost low
4- 8 GHz	Medium to high capacity (200 Mbps to 10 Gbps) 50 km hops typical but ultra-long possible Line-of-sight usually required Spectrum cost can be high, especially in urban areas, due to scarcity
10/11 GHz	Medium to high capacity (200 Mbps to 10 Gbps) Hop distances are location specific and limited by rain and multipath Long hops (>30 km) are possible Spectrum cost high but often less than 4-8 GHz bands

13/15 GHz	Medium to high capacity (200 Mbps to 10 Gbps)
	Hop distances are location specific and mainly limited by rain with some multipath
	Hops up to 30 km are typical
	Spectrum cost less than lower frequency bands but often higher than higher frequency bands
18 - 42 GHz	Medium to high capacity (200 Mbps to 10 Gbps)
	Hop distances are location specific and dependent on rain rates only
	Hop distances are typically less than 15 km (18 GHz) and decrease as frequency band increases
	At 38 GHz, hop lengths typically less than 5 km
70/80 GHz	Very high capacity possible (> 10 Gbps)
	Hop distances are location specific and only dependent on rain rates
	Hop distances are typically less than 2.5 km
	Spectrum costs are often less than lower frequency mmW bands and are often self-coordinated
Future bands above 100 GHz	Ultra-high capacity possible (100 Gbps+) but with short distances (2 km or less)
	Longer distances possible with multi-band and diversity solutions

Unlicensed frequency bands	Planning Considerations
5.x GHz	Capacity is strongly dependent on the level of interference and the finite number of channel bandwidths available on the equipment
	Hop lengths can exceed 50 km
	If PMP, then latency can be an issue
60 GHz	High capacity > 1 Gbps
	<500 m (depends on rain region)

This freedom provides the opportunity for Operators to aggregate parallel links in a configuration called Band and Carrier Aggregation (BCA). Using BCA allows radio planners the opportunity to stretch radio links far beyond their normal link length limits, for a given availability level. BCA will be discussed in more detail in Chapter 7 (Hardware Considerations). Planning assumptions are helpful to set a high-level budget for a project, but for the actual design itself, a detailed design is required as discussed in detail in Chapter 8 (Link Design).

SITE ACQUISITION AND BUILD

Once the customer and repeater sites have been identified, approval needs to be obtained to establish a link. Gaining approval is often the most time-consuming and risky part of the link design process. It is imperative to have a few alternatives in place to minimise these risks.

Planning approvals required include:

- Site owner or site administrator
- Local government authorities
- National planning authorities
- Aviation bodies
- Public and environmental groups
- Wind farm approvals

Tall radio towers and large antennas mounted on rooftops can have a negative visual impact, and therefore, should be minimised or hidden where possible. Plastic trees, water towers, electricity pylons, and even church spires or minarets have been used to lessen the environmental impact. Despite the minuscule safety risk of microwave signals, concerns about RF radiation can also have a detrimental effect on site acquisition, and so this aspect should be considered early on in the process.

The main mitigation against site acquisition risks is ALWAYS to have more than one solution available. Even when the alternative may seem more complex or expensive, it is only once the complexities and costs of the main solution are known, that this conclusion can be reached.

When establishing a new site, it must be remembered that many aspects not related to radio equipment must be taken into account. My biggest surprise when I became the manager of a telecoms network, was that my major operational budgeted expenses had nothing to do with microwave radio equipment. The primary costs and maintenance headaches were mainly to do with roads, towers, equipment shelters, power and earthing.

Most microwave sites are built to support telecom networks that demand so-called five nines availability. To achieve this level of availability, the *telecoms infrastructure* needs to be designed for high availability too.

Unrestricted road access is critical to allow maintenance personal quick access to restore services within strict Service Level Agreements (SLA's). Roads can easily be washed away during the rainy season and require regular maintenance to allow easy access to the radio sites. Given the remoteness of many radio sites, this cost can be prohibitive if the road is long. Choosing sites with easy road access can be a primary site selection criteria in many networks. In a major network I designed in South Africa, we deliberately chose new sites that were close to well maintained existing road infrastructure, albeit dirt roads, to reduce the lifecycle costs of maintaining roads. By keeping access short, road maintenance costs can be slashed. In South Africa, we used a patented system called Hyson cells to quickly and inexpensively lay concrete for the access road to the radio site, thus eliminating road maintenance requirements. As an aside, the various innovative design methods discussed in this book were used to overcome the obvious height considerations of radio sites that were built close to roads, as opposed to being established on hilltops.

Primary power is another, often overlooked, consideration. In many networks the most significant impact on network availability is power. I know of at least one major cellular network where radio planners have been told to select sites based on the nearest power feed as the primary site selection consideration, for the same reason as discussed for roads above. Recent innovations in solar and wind power are creating opportunities for more remote site locations in some geographies.

Towers or poles can be unexpectedly expensive for microwave radio sites. The microwave signal is highly directional and therefore requires stability, even in high wind conditions. This stability requirement, as well as the need to climb it to gain access to the antenna and RF equipment, places enormous design constraints on a radio tower or pole, compared to say a lighting pole. Geotechnical studies have to be done on the soil in the exact location the mast or pole will be built, and in some cases, the concrete footing can cost as much as the pole or tower itself. The physical size of these radio towers can be daunting to radio planners who are used to an office design environment. A 75 m tower on a radio planning design sheet may look innocuous, but I still recall the shock I felt observing the 15 m by 15 m base dimensions of my first large tower that I visited when it was being built. Massive holes had been blasted into the side of the mountain in order to provide space for the reinforced concrete for the tower footing. At a moment like that, you want to be sure that you have got your design right!

Site shelters are the next civil engineering consideration. A concrete base is usually required. Remember that these sites are often in the middle of nowhere, far away from any water supplies, and so can be costly to install. Restrictions on digging foundations can make temporary shelters or shelters that can be installed on a skid, very attractive. Some operators are also building small containerised outdoor shelters that can be placed within the tower footprint or even on the landing on the tower.

Earthing and lightning protection is another important consideration, to eliminate the risk of equipment damage from lightning strikes or surges on the power supply. The best protection for the site is to ensure that everything from the tower and building foundations, to the fence posts and the gate, is bonded together so that there is no electric potential difference between any site components in the case of a lightning strike. A copper strap is often run down the tower, tied to the ground mat, so that cable runs can be earthed. In addition, the cables should be earthed at the point that they enter the building. Additional lightning protection such as a spark arrestor can be added to provide additional protection inside the building.

Other considerations such as **site security** and **environmental impact** should not be overlooked.

New microwave radio site projects are really civil engineering projects, and civil engineering projects are fraught with risks and additional costs. These additional costs can result from several reasons including soil instability, rock outcrops, buried pipes, water seepage, landowner disputes, environmental objections, cultural objections, heritage restrictions, nesting birds or even a rare beetle that cannot be disturbed. The golden rule: plan for alternatives!

REPEATER TYPES

Active repeaters are used to regenerate the signal. For long-haul links, this may include building a new site with costly road access, power supply, a building, and tower structure. In urban areas, an all-outdoor radio mounted on a pole, with an integrated power feed could be all that is required. Examples are shown in Figure 2.3.

An alternative to active repeaters is to use a passive repeater. A passive repeater does not regenerate the signal but bends the signal around the obstruction. Examples of the two main types of passive repeaters are shown in Figure 2.4.

Figure 2.3 All outdoor repeater and long haul repeater

Figure 2.4 Billboard passive repeater and back-to-back antenna repeater

The benefit of passive repeaters is that they are much less costly to build and are more environmentally friendly. They do not require a road, a building, power or a tower and because they are virtually maintenance free, they can be built using a helicopter in virtually inaccessible remote areas, dramatically increasing the possible site options. Due to the physical size of

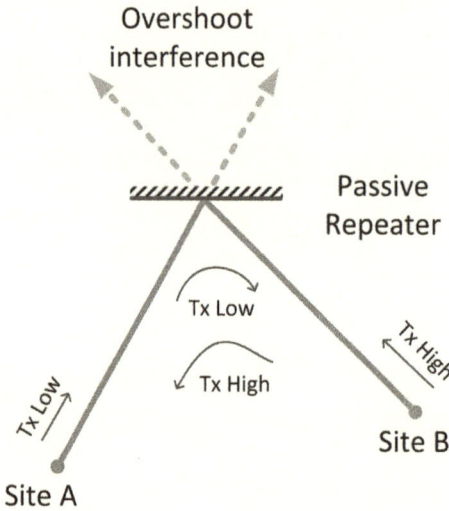

Figure 2.5 Overshoot interference at passive site

a billboard passive repeater, the operating and survival wind speeds need to be taken into account in the design.

A disadvantage of passive repeaters is that they create a secondary interference path due to reflecting the signal. This overshoot interference is illustrated in Figure 2.5. In rural areas where spectrum availability is less scarce, the benefits of a passive repeater may overshadow this frequency disadvantage. In built-up urban areas where spectrum is scarce, overshoot interference is theoretically a problem; however, physical shielding from buildings often eliminates the interference in practice.

The other frequency complication is that at the passive site, a high-low (bucking) clash is created, which is also illustrated in Figure 2.5, and so they are often unpopular with frequency regulators. Bucking is discussed later in Chapter 6 (Frequency Planning). This high-low clash has no practical disadvantage, because usually there is no other equipment at the site. High-low clashes at a passive site, no matter how irrelevant from an engineering perspective, do complicate the frequency administration and is another reason why passive repeaters are often unpopular with, or even disallowed by, frequency Regulators.

The design principle is to calculate the **Insertion Loss** of the passive site and ensure that the link would still meet the performance requirements, often by oversizing the end-site antennas.

$$\text{Insertion Loss (IL)}^{(dB)} = \text{FSL} - (\text{FSL1} + \text{FSL2}) + \text{NPG}^{(dB)}$$

where

FSL = Free space loss between two end sites

FSL1 = Free space loss between Site 1 and passive site

FSL2 = Free space loss between Site 2 and passive site

NPG = Net passive gain in dB

NPG (back-to-back) = A1 + A2 - CL

CL = cable or waveguide loss between the two passives

A1 = Gain of passive antenna 1

A2 = Gain of passive antenna 2

NPG (billboard) = Gain of billboard (obtained from manufacturer data)

Some practical rules of thumb to get a passive site to work, as illustrated in Figures 2.6 to 2.10 are discussed below:

A. Keep one leg short (to minimise insertion loss), in the case of a plane reflector, as shown in Figure 2.6.

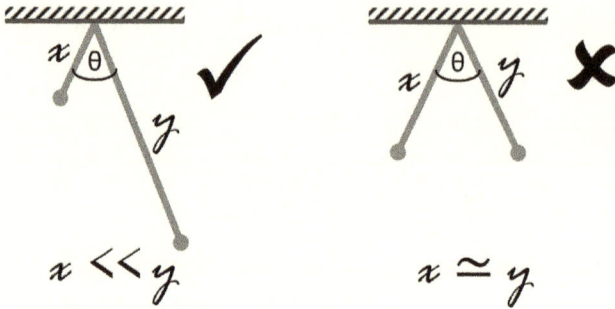

Figure 2.6 Passive link length geometry

B. Keep the included angle between the end sites and the passive site small, to maximise the usable surface area of the plane reflector, as shown in Figure 2.7. The small angle is achieved by choosing a passive site behind the end site.

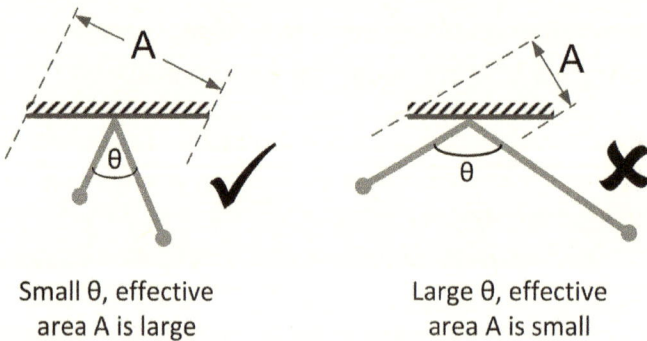

Small θ, effective   Large θ, effective
area A is large   area A is small

Figure 2.7 Passive angle geometry

C. For inline double-reflectors, which consist of two reflectors virtually side-by-side but staggered to face each other and create a net reflection of almost 360 degrees. Ensure that both passives are in the near-field, as shown in Figure 2.8. Near-field and far-field are discussed in more detail in Chapter 5 (Antennas).

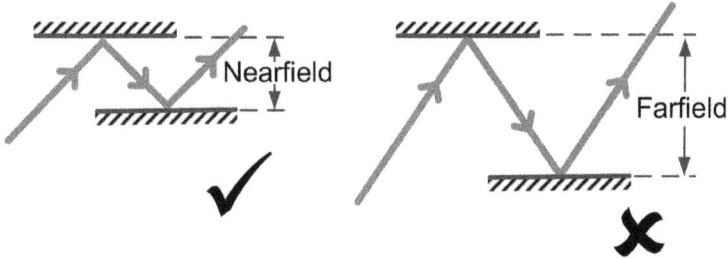

Figure 2.8 Double passive layout

D. Reduce overshoot interference by alternating the polarity of antennas on each path, in the case of a back-to-back antenna system, as shown in Figure 2.9.

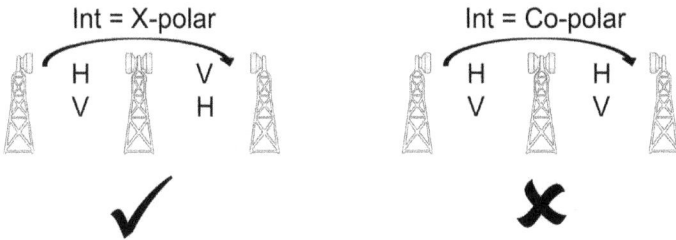

Figure 2.9 Passive link polarisation

E. Ensure that the overshoot signal on a back-to-back antenna path is strongly attenuated (at least 40 dB diffraction loss (DL) even under high k conditions (e.g. k=10), as shown in Figure 2.10. k-factors will be discussed in Chapter 3 (Microwave Propagation).

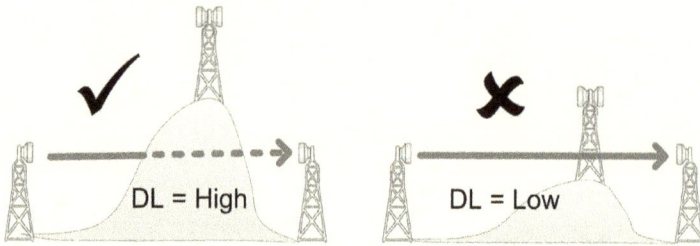

DL = High   DL = Low

Figure 2.10 Passive overshoot diffraction

GIS BASICS

Setting up the radio software for path analysis requires a basic knowledge of the Geographical Information System (GIS) used in the path analysis software program, to ensure that the right settings are used. Analysing paths on a computer with the wrong GIS settings can result in serious errors.

To accurately map a position using software, three main variables need to be considered: The ellipsoid, the datum and the projection method.

The ellipsoid. The earth is not round but is an oblate spheroid. The poles are flattened with the equatorial radius being 21 km greater than the polar radius. The earth is also not symmetrical around the equator, with the south pole being closer to the equator than the north pole. Also, local gravitational variations create minor hills and dales (-100 m to +60 m) that distorts the true shape of the earth from a true spheroid. It is essential to choose an appropriate ellipsoid for the geography being analysed, as if the software has been set up, for example, in the Northern Hemisphere, the errors when applied to the Southern Hemisphere may be many tens of metres.

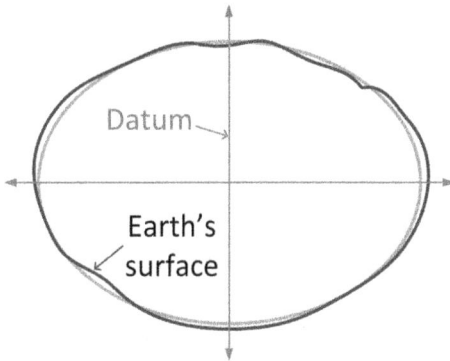

Figure 2.11 The datum

The datum. The reference model that defines where the ellipsoid is placed on the imperfect earth (the geoid) is called the datum as shown in Figure 2.11. Different countries will define a datum that most accurately tracks the actual terrain data for the ellipsoid chosen.

Projection method. The method used to convert spherical coordinates into planar coordinates on a flat surface is called the projection method.

A reliable setting for global applications is the WGS-84 (World Geodetic System) ellipsoid with the WGS-84 datum. Despite its name, it was revised in 2004. Alternatively, the most accurate GIS system that is applicable to the country where the radios are being deployed should be used. For example, in Australia, the GDA-94 (Geocentric Datum of Australia) datum is used.

ESTABLISHING LINE-OF-SIGHT

A starting point to establish a working path is usually to determine if there is, so-called, line-of-sight. Even for *non-line-of-sight* (N-LOS) or *near line-of-sight* radio links, the extent of the blockage of line-of-sight can be useful to know.

In this book, the physics of microwave propagation will be covered in detail. It will be shown that the radio signal is not a pencil-thin, laser-like beam of energy but is, in fact, a wavefront of energy that spreads out into infinity, even with highly directional antennas. All of the energy spread out over these wavefronts contributes to the receiver signal, and so in theory, infinite clearance is needed over any object to get the full receive signal at the other end.

Practically, this is impossible. Fortunately, due to the phase arrangement of the signal across these wavefronts, it turns out that you can still get the full receiver signal even if parts of it are blocked, so long as the blockage occurs at the right phase points. In order to analyse these effects, the exact path details are required. A practical way of doing this is to start by analysing where optical line-of-sight occurs.

The practical use of line-of-sight surveys is in three categories:

For **long hops** over 30 km, the optical line-of-sight information is used to predict multipath fading, reflection fading, diffraction fading and antenna height requirements under a range of refraction conditions. The field survey information is mainly used back in the office for analysis purposes. It should be stressed that the radio trajectory is different from the optical path.

For **short hops**, an optical line-of-sight check is used to ensure that there are no obstacles in the path from the exact location of the antenna bracket at the user end, to the exact location of the antenna bracket at the far end tower. Short hops use high frequencies where refraction effects and diffraction effects are minimal, and so optical line-of-sight is a strong predictor of a good path.

For **non-line-of-sight** paths where N-LOS equipment is used, the survey is often meaningless as the equipment relies on the interference and reflection signals that we are usually trying to avoid. The best case scenario is often where multiple obstacles block the line-of-sight. For example, a non-line-of-sight link in a city where multiple buildings obstruct the path will be likely to work much better than a path obstructed by a single line of trees. For N-LOS equipment, physically installing the link and testing the performance is often a better measure of performance than checking optical line-of-sight. However, there are planning tools on the market that use accurate clutter data to predict performance under non-line-of-sight conditions.

Clearance of the radio signal is modelled using a path profile of the path elevations and terrain clutter, for analysis of the line-of-sight. A path profile is a cross-sectional view of the terrain elevation points between the two ends with the clutter, such as trees and buildings, added on top. Analysis of the path profile can then determine if the radio path has adequate clearance.

Establishing optical line-of-sight in the field is usually achieved with one, or more, of the following methods:

- Binoculars / Telescope
- Light (torch, spotlight, mirror)

- RF source
- Digital or Infrared camera
- Cherry-pickers
- Helium balloons
- Helicopter
- Drones

If optical line-of-sight can be confirmed, the risk of building a radio link over this path is significantly reduced. Establishing line-of-sight is only a starting point for the design, and it should be noted, that optical line-of-sight does not guarantee a link will work reliably, and conversely, the absence of full line-of-sight does not mean a link cannot be made to work. The clarity around optical line-of-sight is used during the design stage where the antenna heights of the link are set. Details of doing an actual design of a radio link are covered in Chapter 8 (Link Design).

Site surveys are essential in planning a radio link to clarify several other practical details, which include the following:

- Accurate site coordinates
- Tower layout and antenna brackets
- Confirming space for equipment
- Cable runs and entry gantries
- Earthing and lightning protection
- Power
- Interference
- Equipment delivery and offload options
- Site access permissions

CHAPTER IN A NUTSHELL

When planning a new link, the starting point is to understand where the connection is required and then determine the exact coordinates of the new site, as well as the required height for the structure that will hold the antenna. The most time-consuming and risky element of the whole radio link design is the site acquisition, and it must always be remembered that building a new site is a *civil* engineering, not *radio* engineering challenge.

Various repeater options are available from all-outdoor to all-indoor solutions and, in some cases, a passive repeater can be built in place of an active one. Two types of passives exist, flat billboard reflector, and back-to-back antenna. Passive sites are cheaper to build and maintain but are unpopular with frequency regulators due to their interference characteristics.

When planning for line-of-sight clearance on a computer the Geographical Information System (GIS) must be set up correctly to ensure the site topology is accurately translated from the real 3D-earth model to the flat models used for analysis purposes. The three variables are the ellipsoid, the datum and the projection model. It is particularly important not to have the computer GIS set up for use in a particular country in the northern hemisphere and then apply it to analysis on link data in the southern hemisphere. The model which is the most generic for any location is WGS-84.

A line-of-sight survey is useful in link planning, to test the likelihood of the link working and establish the required height of the antenna structure. Line-of-sight in itself is not a prerequisite for the link working, nor is the lack of line-of-sight a guarantee that the link will not work, but it is a useful reference point for radio planners to work out the risks involved.

For very short hops, less than 10km, the optical line-of-sight information from the field survey can be used directly for antenna placement, without much further analysis. For longer hops, the difference between the curvature of the radio beam, for different k-factors, compared to the optical line-of-sight signal, is more significant. The line-of-sight information from the survey is thus usually used to increase the accuracy of the path data for analysis purposes, back in the office. Antenna placement on longer hops is done based on clearance rules, which will be discussed in Chapter 8 (Link Design). A line-of-sight survey is not useful for non-line-of-sight (N-LOS) equipment, and a temporary radio link is a preferred method to test for diffraction losses and link performance.

Field surveys are required for far more than just line-of-sight purposes. These surveys are also needed to assess all the practical elements of cost and risk to build the site and link.

Lastly, at the early planning stage, it is useful to establish which radio equipment is most suitable. There is a correlation between hop length, frequency band and capacity as was shown in Figure 2.2.

3

MICROWAVE PROPAGATION

❖ Be able to prove that a microwave signal is not a pencil-thin beam
❖ Correctly identify what a Fresnel zone is
❖ Know in detail what k-factor means
❖ Know what causes diffraction loss and how to avoid it
❖ Know what refraction is and what causes it
❖ Correctly identify when an obstruction such as a tree, electricity conductor, or wind generator blade, could cause problems for a radio path

ANALOGUE PROPAGATION

One aspect of digital microwave systems that has contributed to a poor understanding of propagation issues is that it has been forgotten that the radio signal is still analogue. The digital modulation and processing of the baseband signal results in better transmission performance and is far more robust than historical analogue modulation systems, but the physics of radio transmission has not changed. It is crucial to understand analogue Radio Frequency (RF) transmission in conjunction with the digital baseband processing.

A common misunderstanding in the industry is to assume that the radio signal travels as a highly focussed pencil-thin beam. This erroneous idea crept into mainstream literature when so-called E-band radios (70/80 GHz) were introduced. For the first time, very high gain antennas could be produced with small antennas due to the very small mm-wavelengths. The microwave antenna beamwidth is measured by the angle in which it can focus half the energy into a cone, coming out the front of the antenna. For E-band radio, small antennas that are 30 cm in diameter can achieve a beamwidth of less than 1 degree. One degree drawn on paper is probably the width of a pencil line, which creates the impression that the RF energy is travelling like a laser beam from one antenna to the next. What is forgotten is that large antennas in the lower frequency bands have achieved similar beamwidths for decades, and nobody claimed they were pencil-thin.

Two factors dispel the pencil-thin myth. The first is that if half the energy is focussed inside this cone, the other half is *outside* the cone distributed around the sides and back.

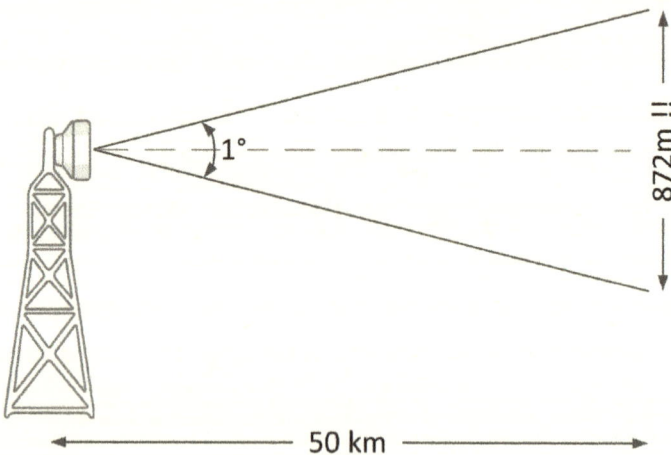

Figure 3.1 Antenna beamwidth

The second factor is that by using simple trigonometry, it can be shown that the diameter of that 1-degree cone is nearly a kilometre, over a 50 km, low frequency hop as illustrated in Figure 3.1.

Later on, it will be shown that thinking about a microwave signal as a laser focussed beam is not a helpful way of thinking about signal density. It is the phase distribution of the energy that matters. To get *any* understanding of the fading issues that impact radio links, to understand how modern Non-Line-Of-Sight (N-LOS) radio links can work, and to ensure that interference signals are correctly analysed for frequency planning, this idea of a pencil-thin beam must be abandoned.

HUYGENS' PRINCIPLE

Physicists studied light signals back in the 1800s and developed various models and mathematical formulas that allow us to understand and predict the behaviour of electromagnetic

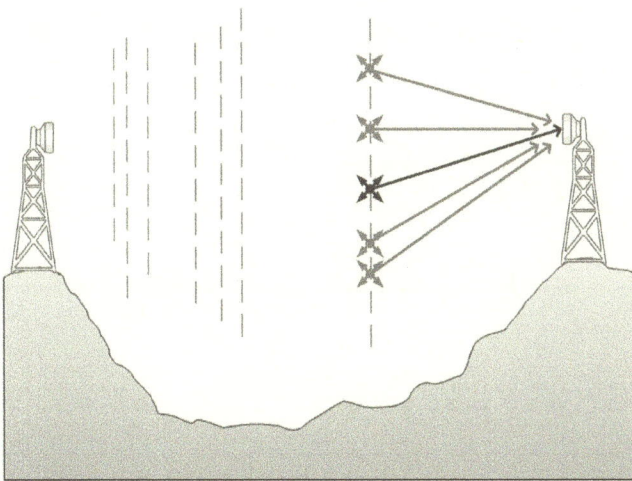

Figure 3.2 Huygens sources

signals when they travel over the atmosphere. These models also help us predict what happens when the signal is partially blocked.

One of those physicists, Christiaan Huygens, developed a model where he suggested that the signal strength at a point can be considered to be the infinite sum of imaginary RF sources that emit energy in all directions, at every point on every wavefront as illustrated in Figure 3.2.

This model is, in fact, the exact opposite of the pencil-thin-beam theory, as it suggests that signals kilometres away from the main beam can theoretically impact signal strength. This model also helps to explain why light can bend around corners, an effect called diffraction. Microwave signals travel over-the-air as an electromagnetic wave, and thus this theory can also be used to predict the signal strength of partially blocked radio signals.

FRESNEL ZONES

French scientist Augustin Fresnel developed some mathematical formulas built on Huygen's theory. Using Fresnel's theory, we can consider the radio signal to be made up of a series of

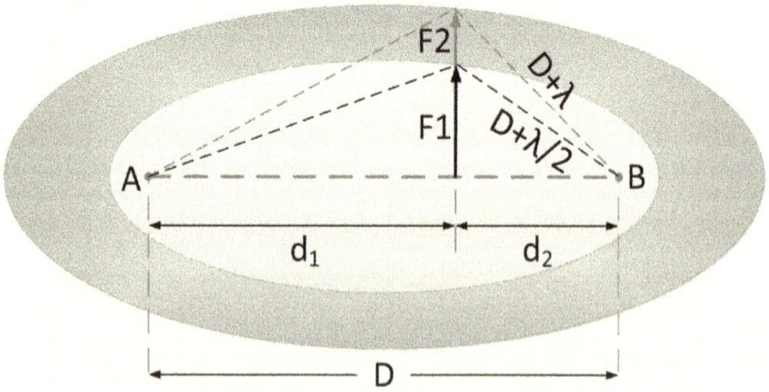

Figure 3.3 Fresnel zones

elliptical zones of energy defined by the Huygens' sources phase contribution to the overall receive signal. It can be seen from Figure 3.3 that the first Fresnel zone is defined by all the points where the Huygen source path lengths are one half a wavelength longer than the direct path. The second Fresnel zone is where the path difference is a full wavelength difference, and so on.

The Fresnel zones are defined in the formula below.

$$F_1 \text{ [m]} = \sqrt{(\lambda (d_1.d_2) / (d_1+d_2))}$$
$$F_n = F_1 \sqrt{n}$$

where n = Fresnel zone number
and d_1, d_2 as shown in Figure 3.3

Fresnel zones are often misunderstood. People sometimes imagine that the radio signal travels in elliptical paths, or is somehow shaped into this pattern. It is incorrect to imagine that somehow the radio rays are travelling in elliptical paths. In reality, Fresnel zones do not exist until some energy is blocked, and have nothing to do with power density. They represent the boundary where Huygens' sources either positively or negatively contribute to overall signal strength due to their phase orientation. Fresnel zones also have nothing to do with antennas as the effect is analysed assuming an isotropic radiator.

The effect of blocking Fresnel zones is illustrated in Figures 3.4 and 3.5. Smooth earth diffraction assumes the rounded earth itself blocks the signal, which is the worst case, as 100 per cent reflection off the earth is assumed. Knife-edge diffraction is the best case as it assumes zero reflections are possible off the obstruction. It can be seen that the worst case clearance condition is where all Fresnel zones beyond the second zone (F2) are blocked, and the effect of negative-phase Huygens' sources is not balanced by positive-phase Huygens' sources located in the odd Fresnel zones (F3, F5, etc.). The best clearance condition is at F1, where blocking the signal produces an enhanced overall receive signal because all negative-phase Huygens' sources are blocked. When 57 per cent - this is often rounded up to 60 per cent - of the First Fresnel zone is blocked, *Clear* the signal strength is identical to a situation where all Fresnel zones are clear of obstructions.

As we will see later, the practical significance is that antennas should not be placed too high, to avoid even-Fresnel zone clearance. They also should not be placed too low, or diffraction loss will occur once less than 60 per cent F1 clearance is achieved. If the diffraction loss exceeds the fade margin, an outage lasting minutes, or even hours, can occur.

In practice, some judgment is required to interpret the risks of diffraction occurring. If an unbroken line of trees growing at a similar height in a forest were blocking the Fresnel zones, it would be a lot easier to analyse than a single branch of a tree waving in and out of the path, or a single electricity conductor crossing the path.

Figure 3.4 Fresnel zone clearance: knife edge obstruction

A practical way to assess the risk of diffraction fading is to compare the physical size of the obstruction, in cross-section, to the size of the first Fresnel zone. Calculating the percentage of blocking just from geometry will provide an idea of how much signal within the first Fresnel zone is getting through.

Let us assume a 10 cm diameter bundled electricity conductor crosses right through the centre of a radio link. Technically speaking, there is no clear line-of-sight and the path is blocked. In practice, if this occurred at the halfway point of an 8 GHz, 50 km link, the percentage of first Fresnel zone area that is blocked is less than 1 per cent and so won't really be noticed. On the other hand, a one-meter wide leafy branch that extends into the path of a 1 km long 70/80 GHz path would block 100 per cent of the first Fresnel zone and so would be unlikely to work.

The position of obstruction is relevant too, even if it does not block the first Fresnel zone. Consider a wind generator that was located precisely at the second or fourth Fresnel distance away from the centre point of the path. The positioning could cause an obstructive loss, and if the blades were reflective (smooth earth at F2), it could have disastrous consequences as shown by Figure 3.5.

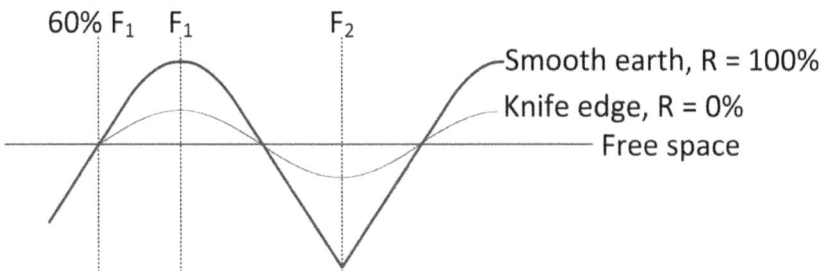

Figure 3.5 Diffraction losses (or gain) as a function of Fresnel zones

REFRACTION EFFECTS

When you turn on the TV channel in the morning to see what the weather forecast is, you probably don't realise that microwave signals are just as interested. Historically, at most airports around the world, weather balloons with pressure (P), temperature (T) and humidity (e) sensors are sent up into the troposphere at midday and midnight to measure the changes in air density - a function of P, T and e - in order to predict what is going to happen to the weather. Most of the fading effects on a microwave radio path result from changes in the layers of air density through which the signal travels.

As already mentioned, the radio signal is not a pencil-thin beam but a wavefront of energy that traverses the path. If the layers of air that the top of the signal experiences, are different from the layers of air at the bottom of the wavefront, bending of the radio signal occurs, resulting in various fading events.

The formula for radio refractive index (N) is shown below, and it can be seen that it is a function of pressure, humidity, and temperature.

$$N = 77.6/T \, (P + 4810 \, e/T)$$

where

T is absolute temperature in Kelvin
P is atmospheric pressure in mbars
e is partial pressure due to water vapour (humidity) in mbars

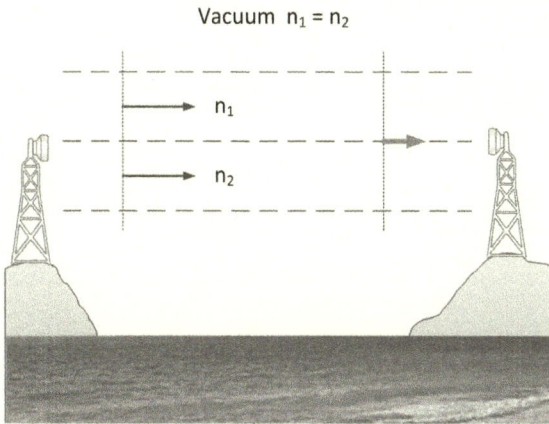

Figure 3.6 Refraction in a vacuum

What radio planners are interested in, is the gradient of refractivity (G) which is calculated by comparing the difference between the refractive index (n_1) at the height of the top of the beam with the refractive index (n_2) at the height of the bottom of the beam, as shown by Figures 3.6 – 3.9. Note that the length of the line indicates the radio ray being advanced or retarded in the differing refractive index densities. The line is longer when the refractive index is lower because the signal travels faster.

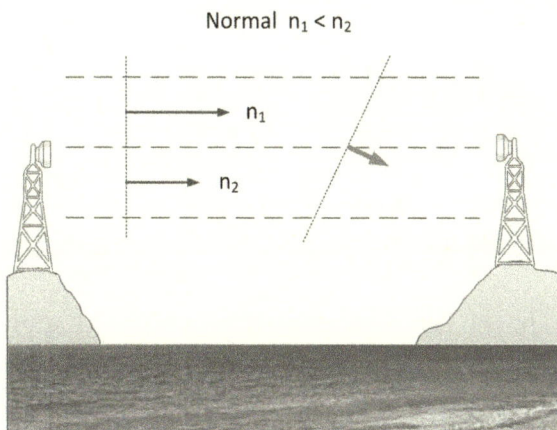

Figure 3.7 Refraction in a normal atmosphere

Abnormal $n_1 > n_2$

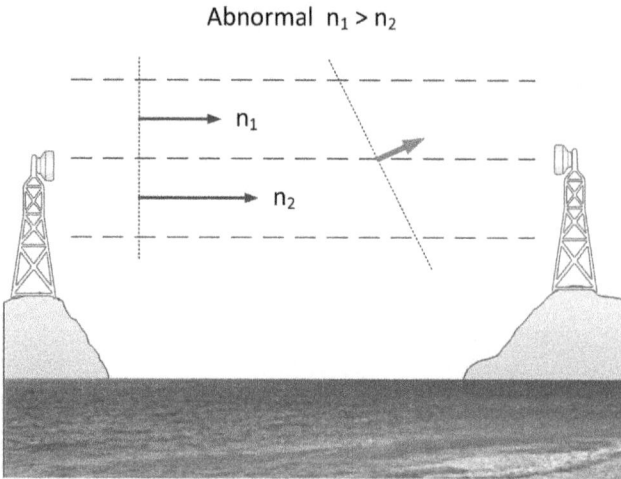

Figure 3.8 Abnormal refraction causing upward bending (possible diffraction)

Abnormal $n_1 \ll n_2$

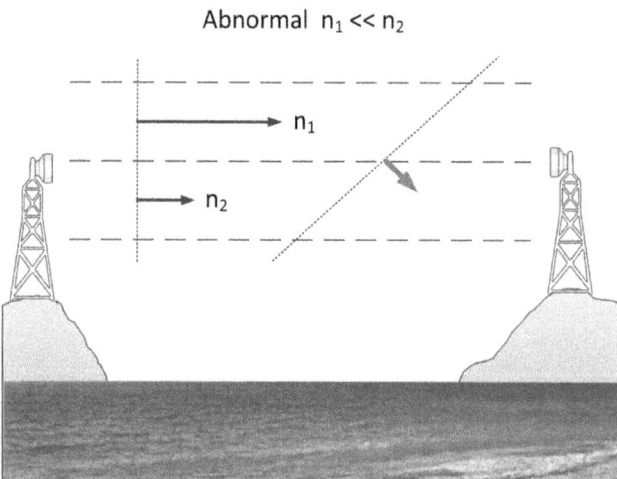

Figure 3.9 Abnormal refraction causing downward bending (possible multipath)

It was mentioned in the Introductory chapter that the speed of microwave signal transmission is a function of the dielectric constant (ε) of the medium it is travelling in. In a vacuum, the signal would travel at the speed of light and, because the top of the beam travels at the same speed as the bottom of the beam, the signal would travel in a straight line.

K-FACTOR

Historically the gradient of refractivity was defined by the so-called k-factor. The k-factor was derived from a method of modifying the earth's radius by a constant - the 'k' factor - to compensate for conveniently drawing the radio beam as a straight line. The method avoided having to analyse the clearance from a curved earth to a curved beam, in the days when a pencil and ruler were the only tools available to radio planners. Curved graph paper adjusted to both the median k (4/3) and minimum k (2/3) were used for analysis purposes.

With personal computers now being used to do these complex calculations, it is odd that this approach is still the standard method programmed into most software analysis tools. Most planning tools still draw the beam as a straight line and falsely modify the earth's radius to compensate. It would be a lot less confusing if the software just modelled what was *actually* happening, with a curved beam over a curved earth.

Confusion often arises when analysing paths modelled with the k-factor method, as what is shown on the computer, while correct regarding *clearances*, is drawn with the exact opposite trajectories to that which occurs in the real world.

k-factor can be calculated from G (refractivity gradient) using the following formula:

$$k = 157 / (157 + G)$$

For most of the time, the air density decreases as the altitude increases and therefore, the signal experiences *thinner* air at the top of the beam. The signal at the top thus travels faster than the signal at the bottom of the beam. As the direction of propagation is always orthogonal to the plane of constant phase, the signal is refracted downwards, which defines the normal path trajectory, as shown in Figure 3.10. When the link is set up, the antennas will be panned slightly upwards to accommodate the downward bending that occurs over the path.

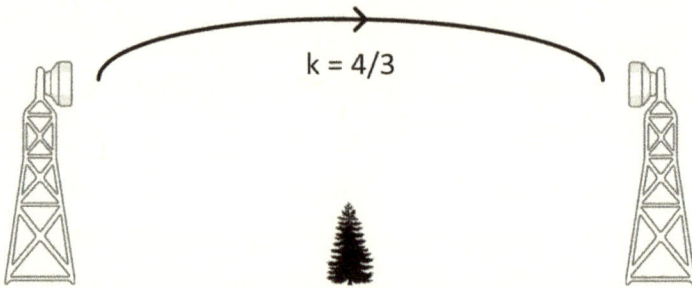

Figure 3.10 Normal k-factor path trajectory

For a small percentage of the time, the air density may vary from its usual condition, thus changing the path trajectory of the radio wave. For example, during a temperature inversion, the temperature may increase with altitude in the lower layers of air. The humidity gradient can vary significantly too and also affects the path trajectory. When the k-factor goes low, the signal is bent upwards, and thus the centre point of the beam will travel closer to any obstructions in the path, as shown in Figure 3.11.

When the k-factor goes high, the signal is bent strongly downwards, and multipath fading can occur, either due to a ground-based or elevated duct, as illustrated in Figure 3.12.

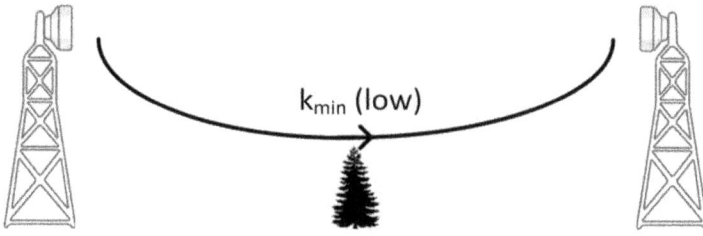

Figure 3.11 Low k-factor path trajectory

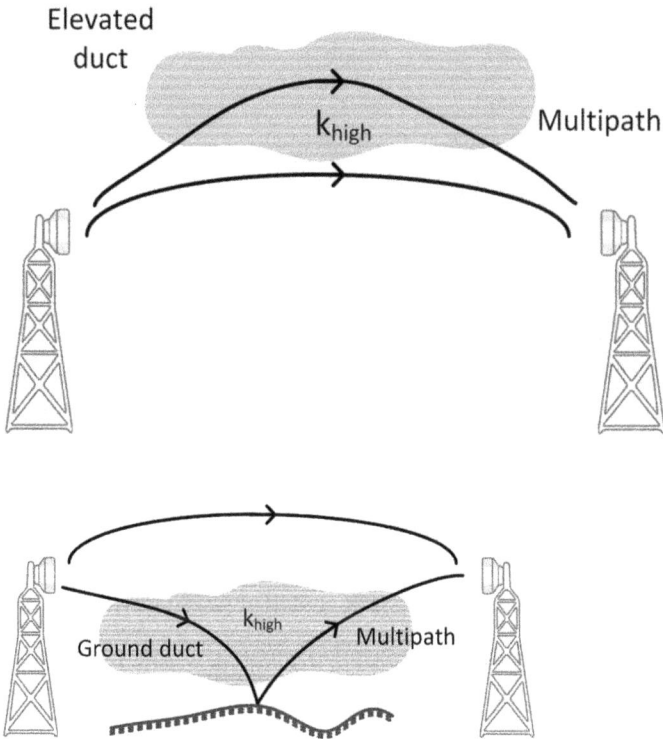

Figure 3.12 High k-factor path trajectory

When the k-factor goes negative, due to extreme ducting, a radio hole can occur with no signal being received at the receiving antenna, as shown in Figure 3.13.

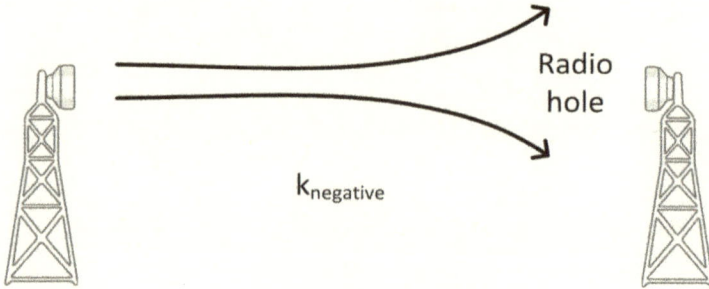

Figure 3.13 Negative k-factor path trajectory

As mentioned earlier, it is the Gradient of refractivity (G) that causes the radio wave refractive bending. Knowing how often the gradient changes and also the depth of change, allows planners to predict the level of fading expected on a path.

Predicting the gradient of refractivity and duct formations is complex. Simplifications are usually made so that some practical conclusions can be made, but it should not be forgotten that these simplifications are not, in fact, accurate. For example, a refractivity gradient or k-factor is usually assumed for a path, but in practice, the instantaneous k-factor at each point along the path is, in fact, different as the temperature and humidity values are not uniform along the whole path, even in stable conditions.

Usually, a high k-factor is caused by hot conditions on a summer morning where dew on the ground, combined with the heat of the rising sun creates a layer of air that is abnormally hot and humid. The same gradient can be created on a dry winter morning where even a small amount of humidity can create a cold layer of air with an extreme humidity gradient. It is the combination of *temperature gradients* and *humidity gradients* as well as the wind-conditions that determined the nature and size of the duct. In addition, while ducts are most common in the morning or at dusk because the earth heats and cools at a different rate to the layers of air above the earth, in practice semi-stable ducts have been observed at other times too including midnight.

CHAPTER IN A NUTSHELL

A microwave signal travels over the air as an electromagnetic wavefront. When the air density closer to the ground is different from the air density higher up, the upper portion of the wavefront experiences different conditions to the lower portion of the beam. This so-called refraction causes the radio signal to have a different level of clearance over the ground when the refractive index gradient changes. The refractive index gradient changes on a diurnal basis as the sun rises and sets. The refractive index gradient also varies seasonally over time, due to different temperature and humidity conditions.

When the signal travels closer to the ground than normal, in other words, when the k-factor is reduced, extra diffraction of the signal can occur. If the diffraction loss exceeds the fade margin of the link, this can result in an availability outage of a few minutes or even hours. In chapter 8 (Link Design), it will be shown how to set the antenna height to avoid this risk.

Conversely, when the k-factor increases, this excessive clearance can result in multipath conditions. Multipath impacts the signal quality due to an accumulation of very short duration signal fading events. A reduction in the antenna height can reduce this type of fading.

In practice, a good design optimises the antenna height to achieve an acceptable amount of diffraction loss while minimising the multipath fading.

4

FADING EFFECTS

- ❖ Identify the types of fading and what causes them
- ❖ Know how to predict a multipath outage
- ❖ Know what the main variables are that impact multipath, so that you can optimise the path
- ❖ Know the difference between flat and selective fading
- ❖ Identify the causes and countermeasures of a specular reflection including the application of space diversity
- ❖ Understand how frequency band impacts rain and multipath fading

TYPES OF FADING

Microwave radio links are impacted by the weather, but not in the way most people think. A heavy rainstorm can cause an outage on a radio link but, rain in itself seldom affects the quality of the link at all in a digital system. Similarly, while abnormal temperature and humidity conditions can create fading problems with additional bit errors occurring periodically on long radio links, these conditions seldom result in an actual outage of the radio signal.

In simplistic terms, the reason for this is that when the signal level falls - or fades - the demodulator can still correctly decipher the data bit as a zero or a one, even with reduced signal voltage. It is only when a critical threshold is reached under severe fading that any errors occur. The receiver threshold characteristics are described in detail in Chapter 8 (Link Design).

The word fading refers to a reduction in receive signal level, due to a variety of reasons. Fading also refers to any condition that causes periodic errors due to signal distortion, such as dispersive fading, where there is not necessarily a signal reduction.

Microwave radios in the millimetre wave bands have fading problems that are completely different from the fading issues experienced on long links in the lower centimetric frequency bands.

Most types of fading relate to refraction effects within atmospheric layers of air, often referred to as ducts. Rain fading is very different and has nothing to do with atmospheric ducts.

It should be pointed out that *any* layer of air is often referred to as a duct, but *true duct* fading, or blackout fading, is when the layer of air has a refractivity Gradient (G) exceeding -157 (negative k-factor).

Diffraction fading is caused by low k-factors, as was shown in Figure 3.11. If the bending of the radio beam results in less than 60 per cent clearance of the first Fresnel Zone (F1), a diffraction loss can occur. In digital radio, this does not necessarily have any impact on quality but, in extreme cases, if it exceeds the fade margin, an outage can occur.

Multipath fading is caused by an unusually high k-factor, as was shown in Figure 3.12. From ray tracing analysis, a k-value above 4 (G<-100) is when multipath conditions are possible from these secondary rays.

Blackout fading occurs when the bending of the radio wave exceeds the real radius of the earth. This negative k-factor results in true ducting conditions and beam spreading of the signal can occur, as was illustrated in Figure 3.13, which can result in long and semi-stable outages of the radio signal.

Reflection fading occurs when a significant portion of the radio signal is reflected off a surface and arrives out of phase, causing an outage. Changing k-factors alter the rays path trajectory, and hence path geometry, thus impacting the reflection fading outages. The effect of path geometry will be covered in detail in this chapter.

Rain fading is the last version of fading covered and has nothing to do with k-factor.

DIFFRACTION FADING

As already mentioned, when discussing Fresnel zones, a diffraction loss can occur if the antennas are set too high or set too low. This fact has been well known for decades yet is often forgotten by today's planning engineers with too much emphasis placed on the risk of too little clearance. I first became aware of this in a practical sense when I was doing some radio planning of long links through a desert-like area in South Africa. This portion of the project was to upgrade old analogue links with new higher capacity digital links and, we were checking path suitability by doing a physical path survey.

With the excellent visibility that was afforded by a pollution-less environment, I could clearly see the opposite end of a 65 km path, from ground level, through my telescope. At both ends of the link, 60 m radio towers had been built. When I considered that for the majority of the time the radio signal has better visibility than optical line-of-sight, because light has a k-factor of 1.15 compared to an average of 1.33 for a radio link, I assumed that the original planners had made an error. Upon further investigation, I realised that the only reason for these giant towers was for 0.01 per cent of the time that the k-factor went to a low value, the so-called minimum k. It dawned on me that for the majority of the time, the antennas were therefore placed far higher than the recommended clearance. In my opinion, many networks perform poorly as the antennas are too high, rather than too low. Clearance rules will be covered in Chapter 8 (Link Design).

MULTIPATH FADING

Multipath conditions occur when the air experienced by the upper parts of the beam is unusually *thin*. The signal trapped in this duct is bent strongly downwards, resulting in additional non-specular reflections, from scattering off the ground. Multipath is also possible from an elevated duct. Ducts tend to occur in the early morning or late afternoon and are often visible because dust, pollution or mist gets trapped in them.

The negative refractivity duct causes the median signal level to be slightly depressed, often through angle-of-arrival fading into the antenna. These scattered signals also cause short, sharp reductions - or enhancements - of signal level, depending on the phase of the interfering signal.

Multipath fading has a very short duration and therefore does not impact signal availability - which by definition is an outage exceeding 10 seconds. This type of fading is mainly a problem for long hops over 30 km, and it will be seen later that the flat fading component increases roughly by the cube of the hop distance.

Multipath fading impacts the link quality in two ways. Firstly, the reduction in signal receive level can cause an outage due to excessive noise in the demodulator. Secondly, the signal distortion that is created across the bandwidth of the receiver can result in an outage, even with a strong receive level present. These two fading mechanisms are called thermal and dispersive fading respectively.

Thermal fading is treated as *flat fading* because it is only the receive signal level that impacts quality. The receiver threshold is a function of the receiver noise floor, which is proportional to temperature as part of kTB - hence the term thermal fading. kTB and the receiver threshold will be covered in more detail in Chapter 8 (Link Design). Antenna sizes need to be large enough to provide sufficient system gain to achieve the right Flat Fade Margin (FFM).

Radio designers can predict the inherent level of flat fading based on the variables that form part of the Po value - called the *multipath fading occurrence factor*. The Po value will be covered in depth in Chapter 8 (Link Design), but the variables that have the most significant impact are the hop length, the frequency, the path slant, the terrain roughness and the refractivity gradient statistics at that location.

Dispersive fading is also called *selective fading* due to its frequency selective characteristics. The errors do not occur from the signal level reduction, but from InterSymbol Interference (ISI) caused by the signal distortion. Theoretically, the selective fading outage is defined by the amount of fading still occurring when zero flat-fade occurs. Radio manufacturers can improve performance against selective fading by incorporating adaptive equalisers into their receivers. To quantify the amount of equalisation provided, a receiver signature curve is generated in the factory and specified as a Dispersive Fade Margin (DFM) figure, derived from the receiver signature curve. Adaptive equalisation is covered in more detail in Chapter 8 (Link Design), and it should be pointed out that it is usually only relevant for hops over 30 km.

On long hops that are over 30 km, antennas need to be placed higher up the tower to ensure that adequate clearance over any obstacles in the path, such as trees, still occurs even when these abnormal gradients are present. In practice, it is often better to focus on achieving optimal clearance under nominal k conditions - where the signal will sit for most of the time - and quantify the risk of having a temporarily reduced fade margin.

On long hops over 30 km, by following the ITU clearance rules - which are designed to reduce diffraction loss - the link may experience a much worse performance from multipath. The reason for this is that the antennas end up being placed well above the optimal height during normal k conditions, with the clearance being even greater under the conditions that cause multipath fading. Multipath signals from the ground tend to get blocked by the terrain when antennas are set at lower heights.

On short hops, changes in path trajectory from k-factor changes are usually insignificant, and antenna heights can be set using information from the practical optical line-of-sight field measurements.

BLACKOUT FADING

Altitude (m)

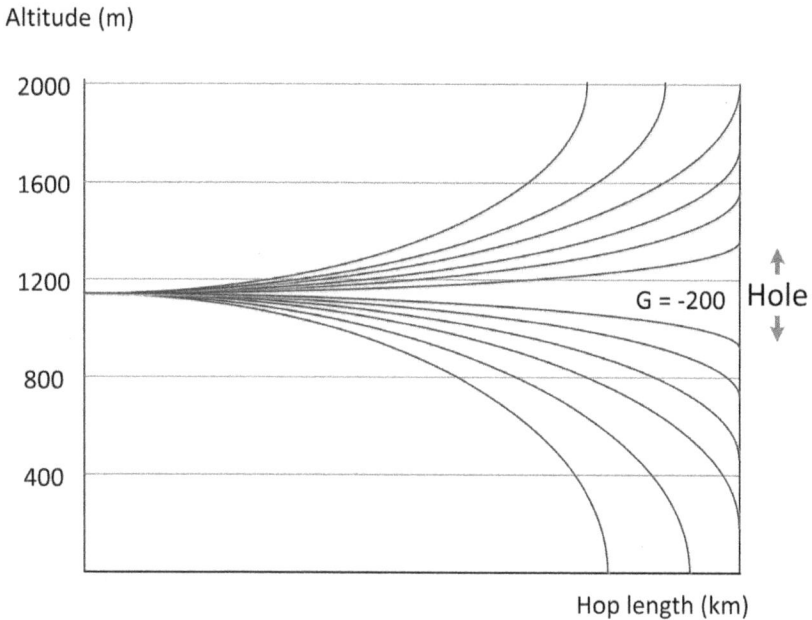

Figure 4.1 Blackout fading with radio hole

When the k-factor goes negative, it means that the radio beam radius exceeds the curvature of the earth. Under this severe ducting condition, the radio signal becomes defocused, and there may be an insufficient signal level for the link to operate. As this is a slow fading event, it is disastrous for the link availability as the link could be down for many hours. A simulation of this type of duct is shown in Figure 4.1 using ray tracing to illustrate how the defocusing occurs. The only way to mitigate this risk is to build very tall towers and put one antenna on the top of the tower and the other at the bottom, and hope that the duct depth is less than this spacing so that one of the antennas can pick up some signal. Fortunately, only a limited number of countries experience these extreme gradients. Usually, they are in hot desert areas, often near the equator.

REFLECTION FADING

In multipath fading, scattered reflections from the ground can create interference. However, they are not stable. When a significant portion of the signal is reflected off, for example, water, the stable interference is called a specular reflection. As a rule of thumb, a reflection will be specular if the smooth surface area is at least the size of the first Fresnel zone. Reflections tend to be worse at lower frequencies as the wavelength is longer, and hence the surface appears more smooth. Vertically polarised signals also perform better than horizontally polarised signals.

Reflection fading is strongly linked to path geometry, as shown in Figure 4.2. According to the law of physics, the angle of the incident ray is identical to the angle of the reflected ray at the point of reflection. When the k-factor changes or where the reflection point moves, such as in tidal variation, the geometry is altered.

Analysing the phase impact of these variations is the key to reducing the effect of reflection fading. For example, the antenna can be placed at the optimal height on the tower - where the reflection is at a minimum - under median k conditions. Placing the antenna at an optimal height will ensure that the

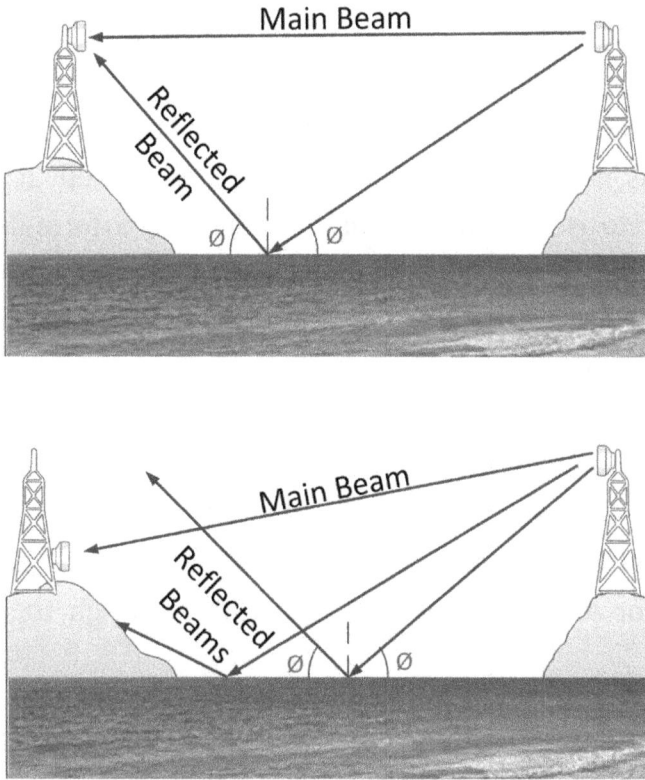

Figure 4.2 Specular reflection geometry

impact of the reflection is at its lowest for the majority of the time. Another technique is to adjust the slant of the path so that the reflection point is moved to a position where it would be blocked by the terrain or clutter such as a building, as shown in the bottom of Figure 4.2. By placing the antenna further back on a rooftop, or choosing a site location where the reflected signal can be blocked by rough terrain, the specular reflection can often be eliminated even without space diversity.

The final and most effective method to overcome reflections is to deploy space diversity, where the two antennas are spaced so that the maxima and minima are at their extreme values. The formulas to determine the improvement from space diversity are complex with many variables including relative permittivity, earth conductivity, grazing angle, and k-factor. Where the reflection surface is not stable, such as water linked to tidal variation, this needs to be taken into account as well. It is recommended to use software that has the ITU-R P.530 formulas programmed into it, to model the path length differences from the path geometry which will indicate the vertical positions on the tower where the signal is at a maximum and a minimum respectively, as shown by Figure 4.3. For a fixed value of k, and a stable reflection point, the path geometry will be such that the combined signal changes with the height on the tower. As the reflection point changes, for example from tidal variations, and as the k-factor varies, the vertical position at which these

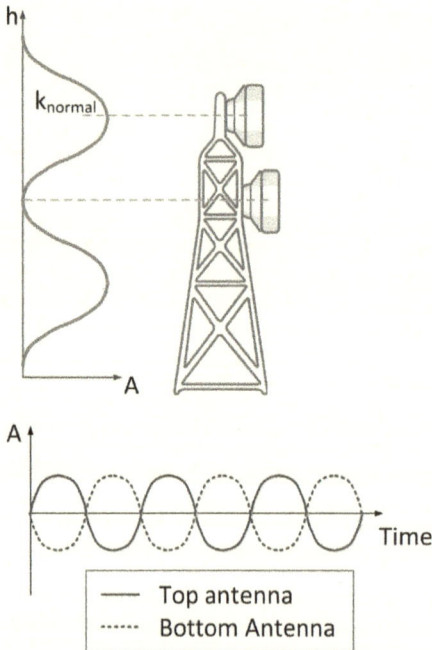

Figure 4.3 Space diversity operation

maxima and minima occur will change. Using this information, an optimum antenna height and separation can be determined. The concept is to place the antenna at an optimum height and spacing so that a fade on the top antenna happens at a different time to that same fade on the lower antenna. In the bottom half of Figure 4.3, the two Receive Signal Level (RSL) plots over time are illustrated for a typical specular reflection. By doing a series of calculations, repeated for different reflection point geometries, as well as modelling a range of values of k-factor, the ideal heights can then be chosen accordingly.

It should be noted that using space diversity for multipath protection does not require a specific spacing as the secondary signal is not stable and is non-specular. In other words, analysing the precise geometry will not help to determine the outage. As a rule of thumb, the spacing should be at least 200 wavelengths to ensure that the signals into each antenna are not correlated. For the typical long-haul bands where space diversity is useful against multipath fading, the required spacing usually works out to between 10-15 m. Providing *less* spacing will still work but may not provide optimum protection. Providing *increased* spacing only improves thermal fading.

With selective fading, improvements are made with closer spacing. Having the lower antenna partially obstructed can improve multipath and specular reflection performance too, due to less reflective signal receiving the dish - recall from Chapter 3 (Fading effects) that multipath conditions are worse with higher k-factor, where there is excessive clearance.

Space diversity improvement is achieved by either switching between the best signal from either antenna, processed at baseband level, or by combining the two signals at the

Intermediate Frequency (IF) level. Space diversity protection can provide significant improvements against multipath fading, and when used with the specific spacing discussed above, it can often eliminate reflection fades.

RAIN FADING

Rain can affect all radio links through absorption; however, for most rainfall intensities occurring outside tropical areas, it is only above 10 GHz that rain has a significant impact. Rain is the primary fading variable when considering link design for the millimetre wave bands. Microwave absorption losses are based on physics, but the absorption curve is not linear with two main resonant peaks at 23 GHz and 60 GHz, as shown by the graph in Figure 4.4 (not drawn to scale).

The resonant peak at 60 GHz limits usable hop lengths to less than 500 m in most places, even ignoring rain, due to

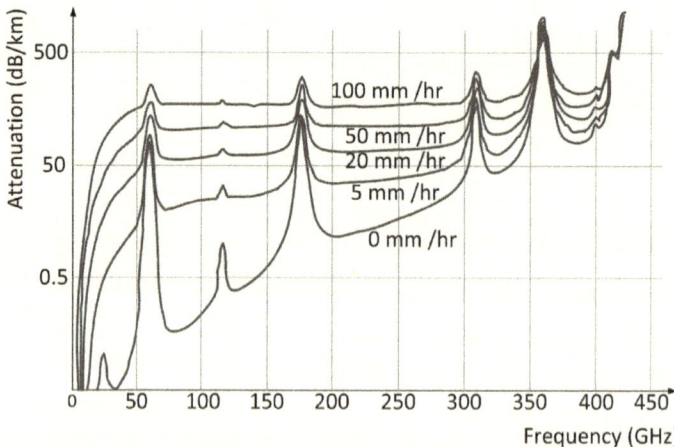

Figure 4.4 Absorption curve as a function of frequency (indicative only)*

* Actual curve obtained from ITU-R P.676

oxygen absorption. The other resonant peak at 23 GHz was chosen by frequency allocation authorities to increase frequency reuse and limits usable hop lengths to around 5 km to 10 km in temperate climates.

In practice, the *location* of the microwave link determines the usable hop length. I have lived in both Johannesburg and London. The stereotypical tourist impression is that in Johannesburg it never rains yet in London it is always raining. The facts show a different story. Regarding total rainfall per annum, Johannesburg is very similar to London (approx 600 mm/annum). Statistically, the number of rainfall days per annum is not that much different either (approx 100). Officially, the *rate* of rainfall is also similar. The International Telecommunication Union (ITU) rain region of both Johannesburg and London is region E. In other words, it doesn't rain harder than 22 mm per hour, more often than 0.01 per cent of the time. The real difference is that in Johannesburg there is almost no rain in winter, but Johannesburg makes up for its rain on summer days. In London, the rain distribution is similar throughout the year. Statistics can be misleading.

What does this have to do with radio link design? Two things: Firstly, check the data to design the links because results vary from city to city. Secondly, check what the data, and the analysis formulae, are based on, as they may not be valid for the specific link you are designing.

We know that when the rainfall rate exceeds the fade margin, the link will fade. Microwave planners will do well, therefore, to take local knowledge of rainfall rates into account. The United Kingdom has published more precise rain maps than those published by the ITU, and the Crane data is more accurate in

Canada and the United States. The 'factual' ITU data is taken from high-velocity tipping-bucket weather stations and then extrapolated. The fewer locations for these primary weather stations, the less accurate the data. In some parts of the world, an entire country's rain region could be classified based on a handful of accurate rain gauges. The attenuation of the microwave signal inside a raindrop is predicted by science, but the computer-generated predictions can often be very different in reality due to limited, accurate rain data sets. The other practical consideration to keep in mind is that when it is not raining, these millimetre wave radio links can be made to work over much longer distances. The predictions are all built on statistical analysis. In other words, how often will the rainfall be hard enough to attenuate the microwave signal beyond the fade margin? Usually, for more than 99.99 per cent of the time, the links could operate over far longer distances!

One common misunderstanding for licenced microwave bands such as 13, 15, 18, 23 and 38 GHz, is to confuse rain-limited hop lengths with usable RF signal level. While atmospheric attenuation above 10 GHz is much higher than below it, it is the rain attenuation that causes the dramatically reduced hop lengths, not the hop length itself. To achieve carrier-grade performance (better than 99.99 per cent availability), during intense rainstorms, hop lengths are decreased accordingly. From an interference point-of-view, the RF signal from these rain affected bands can be quite significant even 50 km away, during non-rainy periods. Table 4.1 illustrates the point that it is only 60 GHz, which 'does not go very far'. The other bands will have a significant signal level at a considerable distance away, from an interference perspective unless physically blocked.

Table 4.1

Atmospheric attenuation

Frequency band	Atmospheric attenuation A (dB/km)	Atmospheric loss over 50 km
18 GHz	0.1	5 dB
23 GHz	0.3	15 dB
38 GHz	0.15	7.5 dB
60 GHz	15	750 dB
70/80 GHz	0.4	20 dB

Regarding 60 GHz, planners need to be careful about the exact frequency that the equipment operates on. The band is relatively wide (57 - 71 GHz), and the atmospheric attenuation is not nearly so high off the resonant peak, which can lead to interference issues, but which can also therefore, make the above technique viable.

BCA (Band Carrier Aggregation), which will be discussed in more detail in Chapter 7 under Aggregation, provides improved performance by exploiting the non-correlated nature of fading events in different frequency bands. For example, an unlicensed 5.8 GHz link could be aggregated with a licensed 15 GHz link to combine the distance, and rain-unaffected benefits of the 5.8 GHz link with the interference-free benefits of the 15 GHz licensed link.

Capacity can also be increased, over longer distances. Using the insights gained in the discussion above, an E-band radio could be stretched over far longer distances than *normal* knowing that, in the event of rain, a back-up 15, 18 or 23 GHz radio - operating on a dual-band feed - would continue to operate due to the lower rain attenuation. I have seen an E-band link operating successfully on a link exceeding 20 km. As discussed, caution must be taken when applying this technique to 60 GHz because rain is not the primary factor that limits the link length, in this case.

This interference issue has mainly been misunderstood in the popular E-band frequency range (70/80 GHz) where it is wrongly assumed that frequency reuse is possible outside a 10 km radius due to signal attenuation. Combined with the misunderstanding about pencil-thin beams, it is predicted that significant problems are likely to hamper these bands as they become more congested, as many planners have ignored the interference challenges. In some countries even the Regulator has misjudged this, resulting in limitations of this band due to interference.

Another misunderstanding in digital radio links is to assume that rain affects the quality level as rain intensity increases. The link runs virtually error-free until the receiver threshold is exceeded and only then does an outage occur. The design methodology is to predict the expected signal reduction during the most intense storms and size the antennas and limit link length to ensure that an outage only occurs for a small percentage of the time. For most applications, no more than 0.01 per cent outage in a year is considered acceptable.

CHAPTER IN A NUTSHELL

Microwave radio links can be affected by the weather, and we call these effects fading. In reality, in digital links when shallow fading occurs, there is no impact on the quality of the system at all.

Most types of fading occur due to changes in the refractive index of the air layers as you go up in height. For short radio hops, which are usually deployed in the spectrum above 10 GHz, the primary fading effect is due to rain.

Refractive fading includes diffraction fading which is caused by the radio signal bending closer to an obstruction than usual under so-called low-k conditions. Antenna heights are set by clearance rules based on the expected diffraction that could occur when these low-k conditions are present.

Multipath fading is caused by the opposite refractivity gradient - in other words, the air layers having a high k-factor. Multipath fading has two components: thermal (flat) fading and dispersive (selective) fading. Radio designers can predict the inherent level of multipath fading based on the variables in the Po formula - hop length, frequency, path slant, terrain roughness and the refractivity gradient statistics at that location - discussed in Chapter 8 (Link Design).

The ducts that create conditions for multipath fading can become so extreme that the radio signal itself gets so-called *defocused*, producing a radio hole. When this occurs, link failure can occur. Fortunately, this is a very infrequent occurrence in most geographies and is mainly a risk factor in extremely hot, desert-like locations often located around the equator.

Where the path traverses water or very flat terrain, specular reflections can occur. These secondary reflected signals are worse for lower frequencies due to the longer wavelengths and can be reduced through careful path geometry or space diversity.

Rain fading is the primary type of fading for frequencies above 10 GHz and is particularly severe for mmWave frequencies. The practical reality is that radio hops have to be kept short to ensure that they still can operate through a rainstorm and so, in general, the higher the frequency, the shorter the hop lengths.

5

ANTENNAS

❖ Understand antenna characteristics such as gain, antenna beamwidth, and sidelobes
❖ Know why we use a parabolic reflector for the dish
❖ Know how to read key data from manufacturer data sheets, and perform link budget and interference calculations
❖ Know the essential characteristic of a high-performance antenna
❖ Know how and why we need to pan an antenna
❖ Know what a waveguide is and when to use it, as well as what precautions to take when installing it
❖ Know what a radome is and why we use it

INTRODUCTION

Most people who have used a portable FM wireless system will be aware of the need for an antenna - sometimes called an aerial. When reception is poor, the antenna can be extended to improve the signal and sometimes pointing the antenna in a particular direction improves reception. Annoyingly, when you let go of the antenna, the reception often worsens. The reason is that your body becomes an extension of the antenna system while you are holding onto the antenna. Antennas are devices that focus or concentrate energy in a particular direction to improve signal strength. Antennas are also the interface that converts the world of current flows and voltages into electromagnetic energy and vice versa.

ANTENNA CHARACTERISTICS

The purpose of a microwave antenna is to concentrate energy in the direction of the path. Antennas - of the same size and frequency - have identical characteristics for both transmit and receive signals - a theorem known as reciprocity. Most microwave dishes are still passive devices. With no power socket they are unable to amplify any signals, so why do we talk about the *gain* of the antenna? Gain is a measurement of the shaping of the signal electrically to have a stronger signal in one direction compared to another. It is analogous to squashing a balloon in a particular direction and saying that it is now longer than the original balloon in that direction. This increase in energy in that specific direction is called gain. To specify the gain of an antenna, we always need to specify the direction. Usually, if the direction is not specified the assumption is that it is out the front of the antenna at zero degrees - the so-called boresight of the antenna.

Antenna gain is the ratio between the microwave antenna and a reference antenna. In microwave systems, the reference antenna has identical energy transmitted in all directions, which is called an isotropic radiator. The gain out the front of the antenna is sometimes called the boresight gain. It is important to remember that an antenna has *gain* in all directions and that this will be different from the *boresight gain* (gain at zero degrees), as will be explained later in this chapter.

The **antenna gain** (G) is thus specified relative to the isotropic radiator in units of dBi.

$$G [dBi] = 10 \log (P / Pi)$$

where P is the gain of the antenna and Pi is the unity gain of the isotropic reference antenna.

The feedhorn is manufactured so that it does not illuminate the entire face of the antenna equally. The illumination angle is reduced. Otherwise, it would cause too much spill-over of energy around the back of the dish. On the other hand, if the illumination angle were too small, antenna gain would be compromised. A practical illumination angle compromise that most microwave antenna manufacturers use is 55 per cent.

Converting this formula into the parameters of a parabolic antenna, and allowing for illumination efficiency (55 per cent), the gain can be expressed in terms of **frequency** (in GHz) and **antenna diameter** D (in m) as follows:

$$G [dBi] = 18 + 20 \log (D * f)$$

Microwave antennas achieve significant gain, compared to lower radio-band frequencies, due to the small wavelengths. The goal is to have all the energy coming out the front of the antenna and none out the back and sides, but in practice, this is impossible to achieve. The unwanted energy is called sidelobes and back lobes, as illustrated in Figure 5.1.

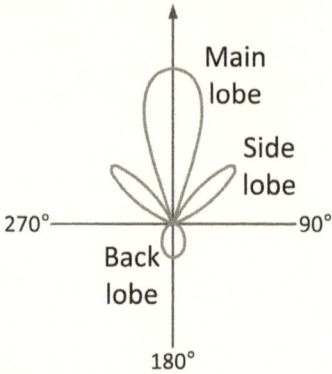

Figure 5.1 Polar antenna pattern

The directivity or gain of an antenna increases when the beamwidth decreases, as shown in Figure 5.2. The beamwidth is more accurately named the half power beamwidth (HPBW), because it is the angle created when the power has dropped by 3 dB, from the peak power at the front of the antenna. A 3 dB drop represents a halving of power.

The formula to convert from gain to **Half Power BeamWidth** (HPBW) is shown below:

$$\text{HPBW} = 163/ \sqrt{(G)} \text{ in degrees}$$

where $G = 10^{G[dB]/10}$

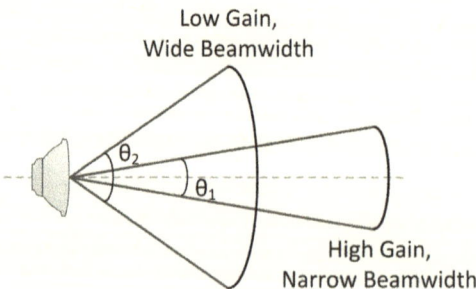

Figure 5.2 Gain to beamwidth relationship

RADIATION PATTERN ENVELOPES

In practice, to know the exact characteristics of an antenna for design purposes, it is necessary to consult the antenna manufacturer's gain curves, called the Radiation Pattern Envelope. (RPE). When doing interference calculations, it is helpful to imagine that the antenna is made up of 360 antennas, each spaced one degree apart around the tower leg. Also, imagine that there are an additional 360 antennas on the opposite polarisation, spaced one degree apart. Each set of antennas, in essence, creates four paths - two co-polar paths (HH, VV), and two cross-polar paths (HV, VH). The RPE curve effectively shows the gain of each of the 360 co-polar antennas, as well as the 360 cross-polar antennas, in the form of co-polar and cross-polar patterns. An example of an RPE curve is shown in Figure 5.3. Note that an expanded scale is used for the first 20 degrees in the example shown, to exaggerate the shape of the main lobe.

In the case of ultra-high-performance antennas, the RPE pattern is often plotted for both sides of the dish (+ 180 and - 180 degrees). The reason for this is that the J-arm of the horn feed creates a non-symmetrical pattern depending on which half of the antenna it is located in. In very critical interference situations, the antenna may be placed on the best side to optimise the interference situation. In new Cassegrain horn feeds this problem is now eliminated, and both antenna halves are symmetrical with optimised performance.

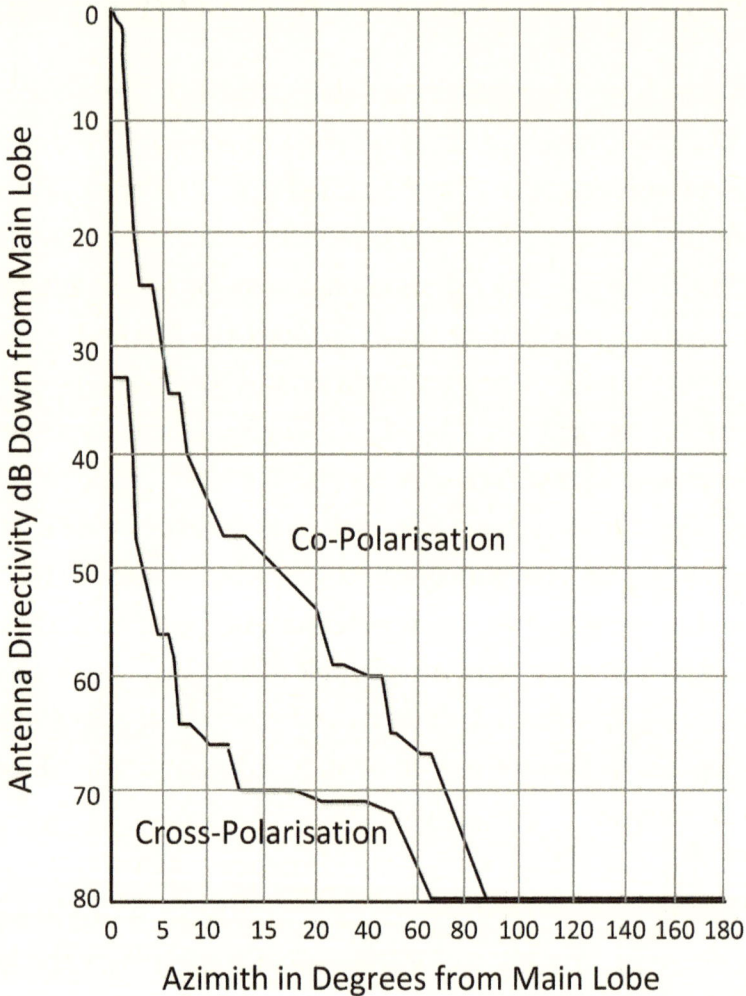

Figure 5.3 Radiation Pattern Envelope

An example of how manufacturers create the mask is shown in Figure 5.4. Both co-polar, as well as cross-polar patterns, are usually provided.

The **boresight gain** is usually specified for each antenna, and the graphs are normalised relative to that figure.

The **beamwidth** can be read off the curve, by locating where the directivity (antenna gain) has fallen by 3 dB.

The **Front-to-Back ratio** (F/B) is the difference between the gain at the front of the dish compared to the gain at the back of the dish. It will be seen in Chapter 6 (Frequency Planning) that a good F/B ratio is needed for efficient frequency reuse.

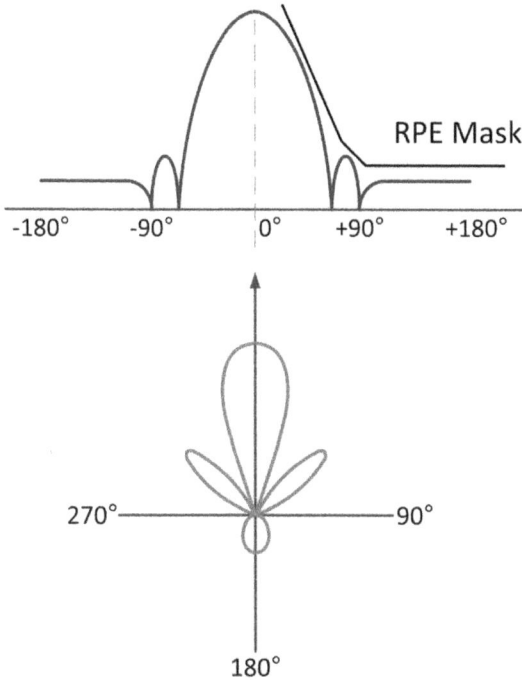

Figure 5.4 Rectangular coordinate conversion from polar pattern

The **Cross-Polar Discrimination** (XPD) is the difference between the gain of the co-polar pattern, compared with the gain of the cross-polar pattern. It can be seen from the Radiation Envelope in Figure 5.3 that XPD is much greater at the front of the dish (0 degrees) compared to the back of the dish (180 degrees), which is an important consideration when doing frequency planning. Using cross-polar operation as an interference-limiting scheme only works if the interference is into the front of the dish, which will be covered in more detail in Chapter 6 (Frequency Planning).

RETURN LOSS (VSWR)

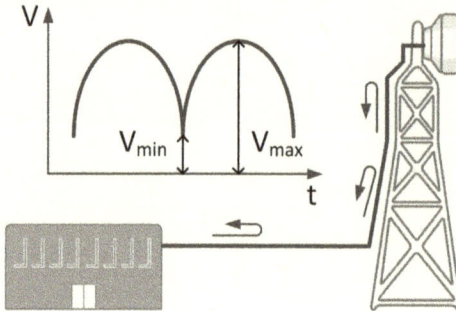

Figure 5.5 Voltage Standing Wave signal generation

When a radio frequency signal is injected into a cable (coaxial or waveguide), that is subsequently connected to an antenna, not all the energy gets converted to electromagnetic energy. Some of it gets reflected back to the transmitter module due to a mismatch in impedance at the connector interfaces, and due to imperfections in the cable, as shown in Figure 5.5.

This reflected power is subsequently re-transmitted at the transmitter interface too, as it is not matched for the reflected signal, which sets up a standing wave in the cable. The **Voltage Standing Wave Ratio** (VSWR) is the difference between the voltage maxima and voltage minima in the cable. The better the matching, the lower the VSWR value, where a value of unity represents a perfect match.

$$VSWR = Vmax / Vmin$$

This value can also be represented as the **Return Loss** [dB] using the following formula:

$$\text{Return Loss [dB]} = 20 \log (1/\rho)$$

where ρ (reflection coefficient) = (VSWR-1)/(VSWR+1)

In the past, return loss was a critical factor as unwanted reflected power could damage the transmitter circuitry. Today the transmitter circuitry could even withstand a short circuit, so even a poor return loss will not damage the equipment. The importance of return loss is related to performance degradation, not equipment damage.

Where waveguides are used, a good return loss is essential, as any physical distortions and impedance mismatches would have a direct impact on link performance due to signal distortion, thus increasing selective fading. Adaptive equalisers often mask the cable problems, so it is good practice to sweep the cable to test the return loss before connecting the traffic. A good field figure to aim for is a VSWR better than 1.4 or 1.5 (return loss approximately 25 dB). Where Intermediate Frequency (IF) or baseband cables are being used, the RF circuitry is located in the OutDoor Unit (ODU) and so Return Loss is not such a critical characteristic.

NEAR-FIELD

When a microwave signal is transmitted, there is a conversion between the circuitry world of current flows and voltages, to the over-the-air electromagnetic propagation world. When the transmitting antenna is far enough separated from the receiving antenna, the two can be treated as two independent entities, and the signal transmission can be analysed using geometric optics, where the ray paths can be analysed to understand their behaviour. Concepts such as Fresnel zones, antenna gain and the formula to determine Free Space Loss (FSL) make this far-field assumption, in determining the formulas.

There is a transition point where the two antennas can no longer be regarded as separate as their fields interact with one another, and this area close to the antenna is called the **near-field.** In the near-field, the receiver signal strength is no longer inversely proportional to distance, and the power density becomes oscillatory.

The **far-field distance**, which is where the signal is no longer in the near-field, is defined by the following formula:

$$\text{Far-Field [m]} = 2\,D^2/\lambda$$
$$= 6.7 \cdot D^2 \cdot f_{GHz}$$

where D = antenna diameter in metres, and f is the frequency in GHz.

Regarding the practical significance of the near-field, there is not much that can be done to predict signal levels, as no formulas have been developed for the near-field, however, at least being aware of where the zone is, is important. Planners often make the mistake of trying to use Fresnel zones to work out the impact of a local obstruction, when Fresnel zones are a far-field concept. Also, when back-to-back antenna systems are being designed, there is a risk of operating inside the near-field zone, and antenna gain can *actually go down* when bigger antennas are used due to close coupling of the dishes.

Historically, a rule of thumb was to keep a 30-degree cone of protection free from near-field obstacles, but in practice, this is hard to do. As a minimum, a cylinder-shaped zone of the same diameter as the antenna should be kept free of obstacles to avoid any near-field reflections.

ANTENNA TYPES

In microwave point-to-point transmission, we still predominantly use parabolic antennas to shape the signal into a plane Transverse ElectroMagnetic (TEM) wavefront. Passive flat plate phase array, and active antennas, with solid-state electronics shaping the RF energy, are usually only financially viable in class licenced (unlicenced) bands where manufacturing volumes are very high. They are also increasingly being used in the millimetre wave bands such as E-band and V-band, where a small flat plate antenna can be designed for low visual impact, while still meeting stringent design standards, such as Class 4 antenna masks. Some manufacturers are using the modern technology used in 5 G, such as beam-forming and applying it to microwave antennas. Waveguide slot antennas are also now being developed for mmWave bands, such as 60 GHz and are expected to be used in future for bands above 100GHz.

What is often misunderstood is that different parabolic antenna types do not differ in gain. The primary consideration is an improved shape of the antenna beam for frequency planning reasons. As will be explained in Chapter 6 (Frequency planning), if the antenna back lobes and sidelobes can be sufficiently reduced, the same frequency channel can be used multiple times.

The principle of a parabolic dish is that if a signal is injected from the focal point and reflected off the parabolic shaped reflector, all the rays end up in phase and parallel, creating a plane wavefront, as can be seen in Figure 5.6. It should be noted that only the horn-feed is frequency specific. Integrated multi-band antennas can be created with a dual frequency horn-feed reflected onto a single antenna reflector.

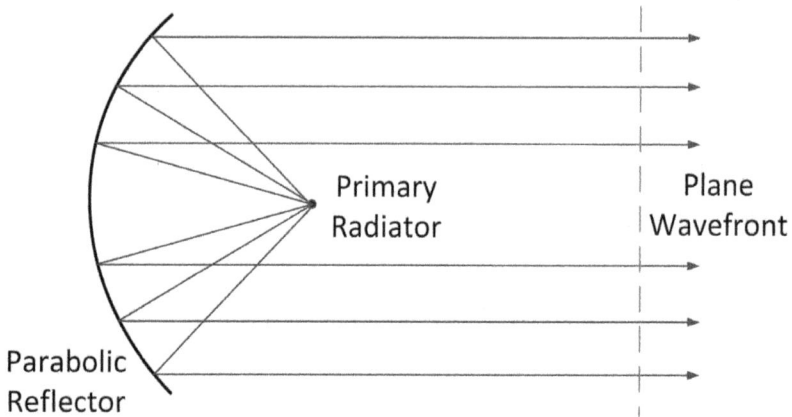

Figure 5.6 Parabolic antenna geometry

A grid antenna uses grid rods to approximate the shape of the antenna. Where the frequency is low enough - usually below about 3 GHz - the electrical performance is identical to a solid dish, providing the grid rods are spaced closely enough relative to the wavelength of the signal. As an example, a 2.4 GHz signal has a wavelength of 12.5 cm so a rod-spacing of 1 cm would be less than a tenth of the wavelength and the grid would behave like a solid reflector. One limitation is that only one polarisation is supported, due to the orientation of the grid rods.

Grid antennas are popular due to the significant cost savings. Apart from a reduction in materials used to manufacture the dish, the wind loading reduction on the tower is substantial, resulting in cost savings on the tower stability required. Also, transportation costs are reduced by as much as 30 per cent for large antennas. In the case of large grid antennas, the antenna can be shipped in a collapsed mode and assembled on site.

Improved Front-to-Back (F/B) performance and sidelobe suppression can be obtained by extending the parabola shape around to the focal plane, and further improved by adding a shroud around the dish, called a High Performance (HP) dish. To avoid reflections off the side of the shroud and the horn feed itself, these elements are covered by carbon-loaded foam to absorb the unwanted signals. The foam adds significant costs to the antenna but helps with frequency reuse. For this reason, regulators will often insist on using these improved HP antennas in congested bands. Hence, antennas are manufactured with a class rating. In the US, they are called category A and B, and in Europe, ETSI specifies four classes of antenna.

It should be noted that higher class antennas can be significantly more expensive. For example, ETSI class 4 antennas can easily cost double a similar sized class 3 antenna. However, the additional hardware cost should be considered in light of the cost of spectrum. Spectrum is a scarce resource and when it is gone, it is gone. It could be considered priceless!

In a network I designed years ago in South Africa, the entire network was built on a single frequency pair, through using high performance antennas, using the frequency reuse techniques discussed in Chapter 6 (Frequency planning). This allowed the capacity to be increased multiple times over the years that this system was in operation. An alternative network did not use frequency reuse, and for the same geographical coverage ran out of spectrum after the first upgrade. It is my assumption that the money saved on *standard* antennas paled into insignificance compared to the cost of such inefficient spectrum use.

Microwave antennas called slip-fit antennas are also manufactured where the radio manufacturers collaborate with the antenna manufacturers to produce a proprietary flange that allows a direct connection from the ODU (OutDoor Unit) to the antenna, without the need for an expensive waveguide connection. Even short waveguide sections also add transmission losses, which effectively reduces the net gain of the antenna.

ANTENNA INSTALLATION ISSUES

Antennas generate a significant loading on a tower structure, especially in high winds or with the additional weight of ice and snow. A structural analysis should always be carried out before installing a dish to ensure the tower can withstand these forces. Larger dishes may need to have side struts that provide further stability.

Microwave antennas must be aligned such that the boresight gain of the antenna is experienced under the median k-factor condition, to ensure optimal performance. Optimal performance is achieved by so-called panning of the dishes, which is shown graphically in Figure 5.7.

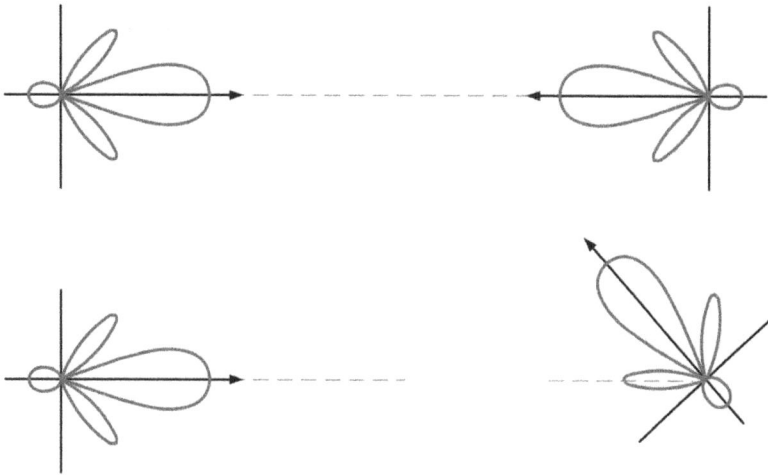

Figure 5.7 Antenna panning

Due to the small beamwidths of microwave antennas, panning must be done through a fine-tuning mechanism and not by large physical movements of the antenna on the pole. One degree of movement is achieved with mm, not cm, of rotation on the pole. As discussed in Chapter 3 (Microwave propagation), the industry has erroneously linked the narrow beamwidth of small E-band antennas with the high frequency (70 / 80 GHz) rather than the real reason, which is the beamwidth. A large antenna at 7 or 8 GHz will have a very similar beamwidth to a small (30cm) E-band radio.

It is advisable to calculate the azimuth of the link and set the antenna up at the correct bearing before panning. For the majority of links, the elevation of the antenna will be close to level (zero degrees), with only a fractional uptilt for k-factor alignment. Even where there are large height differences between the two end sites, the inclination is small, as the vertical height difference is typically 10's or 100's of metres, whereas the horizontal distance is in kilometres.

A practical alignment tip is to use a pocket spirit level to ensure that at the start of the panning process, the vertical alignment is at zero degrees uptilt for all links. With the antenna perfectly horizontally aligned in elevation, and the azimuth position pre-estimated, the rigger can move the antenna left and right of this position to find the side lobes and then lock the azimuth on the maximum signal. At this point, fine adjustment of the uptilt can be made to align with the maximum signal, which will usually be associated with path inclination linked to a k-factor of 1.33. By setting the antenna up at the correct azimuth, in addition to having no up or down tilt in the horizontal plane before panning starts, the risk of sidelobe panning, or panning onto a stray reflection, is reduced.

A typical installation error is to align the sidelobe, instead of the main lobe of the dish, which is often done in elevation, even if the azimuth is correctly aligned. In digital systems, the link will still work. However, performance will be degraded due to a reduced fade margin.

Typically, the first dip is approximately 30 dB lower than the boresight peak, and the first sidelobe peak is approximately 20 dB lower than the boresight peak signal. The exact figure is hard to estimate because the actual null value is not shown on an RPE curve. Recall that the RPE curve indicates the maximum level, not the minimum. If these peaks and troughs are not observed as the antenna is rotated into full alignment, it is possible that the antenna is being aligned off-peak. In the lower frequency bands, below 10 GHz, finding a peak signal that is 20 dB too low can be a telltale sign that sidelobe alignment has occurred. Alignment errors are easy to make at higher frequencies, or on larger antennas where the beamwidth can be a fraction of a degree. Another common mistake is to install the link with cross-polarised antennas, resulting in a 30 dB difference in signal level due to cross-polar operation.

ANTENNA ACCESSORIES

A feeder cable is required to connect the radio equipment with the antenna. In the past, the majority of microwave links were installed with waveguides. However, today most equipment has the RF unit installed outdoors with an Intermediate Frequency (IF) or baseband cable connecting the indoor unit. In some cases, the entire radio is outdoors. Where waveguide is used, it needs to be carefully planned to avoid excessive bends or any twisting, and it should be pressurised to keep moisture out. Even a pinprick hole can cause water ingress to build up. I have personally seen a bucket full of water pour out of an unpressurised waveguide when the connector was removed. Even though this was a sub-10 GHz link, the amount of signal absorption from the water caused this link to stop working. Pressurisers and dehumidifiers are required on long waveguide runs, but are not necessary on short runs such as the flexible waveguide that is often used between a tower mounted Outdoor Unit (ODU) and an antenna. Special clamps are required to fasten the cable to the tower in such a manner that the vertical cable weight is supported - to avoid bunching of the copper corrugations. The clamp should not be so tight that it dents the copper corrugations as this would distort the RF signal. Waveguide losses are high, with a typical waveguide cable at 7 GHz exhibiting 6 dB loss over 100 m. This is equivalent to a halving of antenna gain, hence the popularity of split configuration, or all-outdoor equipment.

A protective covering called a radome is usually recommended to reduce the wind loading of the antenna on the tower. Also, it protects the horn-feed from damage that may be caused by things like birds or ice. At higher frequencies, hydrophobic materials may be required for the radome to ensure there is not a significant radome loss during wet weather. Where antennas are installed in extremely icy conditions heating of the radome may be required. AC-powered stick-on pads or clamp-on panels are available in the marketplace to retrofit to standard antennas and, there are also vendors who provide a hydrophobic material that wraps around the antenna to stop the build-up of water or snow thus preventing ice formations. The difficulty is achieving compliance from standards bodies such as the European Telecommunications Standards Institute (ETSI), which can restrict the use of the devices discussed above.

The antenna installation with its cable connection to the equipment room may increase the risk of lightning damage. For this reason, it is good practice to establish a common ground potential through bonding everything together including the site fence, the tower, the building, and the cables. Ideally, the grounding of the cables should be strapped to a flat copper bar at the waveguide entry and taken to ground potential outside the building to avoid any lightning currents inducing voltage spikes inside the equipment room. Some organisations try to get a good earth by driving metallic rods deep into the ground. This strategy of creating a very low resistance ground connection, less than 5 ohms, is hard to achieve on a radio site, as radio sites tend to be built on rocky hilltops, where soil resistivity is high. The alternative approach is to bond everything together, and not worry too much about the ground resistance. In practise, lightning currents have a wide frequency spectrum, and so the impedance is more important than the resistance. By bonding everything together, a site can withstand even a direct strike as everything goes up at the same potential, thus eliminating the damage that can be generated by current flows caused by the electrical potential difference.

CHAPTER IN A NUTSHELL

An antenna is required to concentrate the energy in the direction of the path thus providing *gain,* which increases the usable signal at the other end. Gain is achieved by reducing the antenna beamwidth. This enhanced signal increases the system gain in the link budget and allows a radio link to overcome fading issues so that the link works error-free during rain and atmospheric ducting conditions. The more concentrated the signal is out the front of the antenna, the less energy goes out the back and sides, thus reducing the risk of interference. Signals around the sides of the antenna are called sidelobes. The quantification of the antenna signals from zero to 360 degrees is shown on a Radiation Pattern Envelope (RPE) diagram, which is available from antenna manufacturers.

Most microwave dishes use a parabolic reflector. If a signal is injected from the horn feed, which is located at the focus of the parabolic shape, and then reflected off the parabolic reflector surface, all signals will be parallel and in phase as they have travelled the same distance irrespective of where they were reflected off the parabola. The signal now travels in a straight line, providing it was in free space, as a plane wavefront.

Apart from the gain, which is used to increase the overall system gain of a link, antennas are used to reduce and manage potential interference risks. Improved front-to-back ratios can be achieved by high-performance antennas which reduce side lobes and back lobes, thus improving performance where reusing the same frequency.

A key installation consideration is to ensure that the antennas are panned onto the main lobe and not onto a side lobe or the alternative polarisation.

Where all-indoor radio equipment is used, a waveguide is required to transport the RF signal to the antenna. Waveguides are very expensive and require careful planning and installation as they cannot be twisted and do not tolerate any deformations, even a small dent. Waveguides also absorb a significant amount of RF signal, which is why so many systems today have the RF unit located behind the antenna. An outdoor RF unit eliminates the need for a waveguide or reduces the waveguide requirement to a short flexible connection cable. Antenna radomes are used to reduce wind loading on the tower, and they also protect the horn-feed.

6

FREQUENCY PLANNING

- ❖ What is frequency interference?
- ❖ Who decides which frequencies I can use?
- ❖ What licensing options are available?
- ❖ What does high-low (or A/B) planning mean?
- ❖ When can I reuse the same frequency again?
- ❖ What is threshold degradation?
- ❖ Explain overshoot and nodal interference
- ❖ Know how to plan a network for frequency reuse

INTERFERENCE

Most people have experienced a crackle on their radio or TV sound during an electrical storm. The crackle is caused by stray signals from the increased white noise of the lightning strike breaking through to the receiver. Telecoms receiver equipment will demodulate any signal within its passband, and all signals that do not originate from the original source are considered to be interference. Interference analysis is a specialist field, but the basic issues should be understood to ensure a quality design, as, in digital systems, interference can be a hidden gremlin. People may think their network is interference-free, but in fact, the interference may be present and only become an issue as the link fades. For example, a light rain storm that should have no impact on system performance could result in excessive errors and poor performance due to the latent interference, rather than the rain itself.

In analogue radio networks, interference was a primary contributor to network performance even in unfaded conditions. Any signal degradation had an immediate effect on the system. In digital systems, the equipment is far more robust, and much less sensitive to interference issues. It is usually only in a faded condition that interference is noticed. Many designers only consider Threshold (T) to Interference (I) conditions, as the analysis of the unfaded Carrier (C) is not of concern. Ironically, the robustness of digital systems has often resulted in worse-performing, rather than better-performing networks. The reason for this is that with analogue radio any performance issues were noticed at the time of commissioning in unfaded conditions. It was imperative to resolve them as everyone was aware there was an issue. In digital systems, very high levels of interference may be present but until the system fades there is no noticeable degradation in link performance, and so nothing is done to address the issue. Often, network operators are not even aware that they have this hidden performance issue.

Some equipment provides visibility of the SNR through the ACM system, and in cases of high potential interference, a spectrum analyser can be used at the site to identify rouge signals. The ideal approach to checking for interference during commissioning, if the equipment allows it, is to fade the signal to threshold to check if the threshold is degraded. Failing this, reducing interference through good antenna selection, as discussed in Chapter 5 (Antennas), and deploying proper frequency planning measures, as discussed in Chapter 6 (Frequency planning), should be considered.

The International Telecommunication Union (ITU) regulations define interference as 'The effect of unwanted energy due to emissions, radiations or inductions, manifested by any performance degradation, misinterpretation, or loss of information, that could be extracted in the absence of such unwanted energy'. In other words, it is any unwanted signal that affects the demodulation of the wanted signal.

Interference can come from several causes:

- An external signal from another system on your channel
- An external signal with a harmonic / sideband / spurious emission on your channel
- Leakage from your adjacent channel (filter selectivity)
- Overshoot interference from another link further down the chain
- High/low clash
- A rusty bolt which acts as a diode and generates InterModulation Products (IMP's).

The fundamental design issue is that the demodulator requires a minimum Signal to Noise Ratio (SNR) to operate error free. This SNR value varies depending on the modulation scheme. With complex modulation schemes, this value may be very high, so even a very low-level unwanted signal may degrade the demodulator performance by adding to the noise floor. When the threshold is degraded, the link may not have an adequate *Effective* Fade Margin (EFM), despite having a good

Flat Fade Margin (FFM). The degraded threshold is illustrated in Figure 6.1.

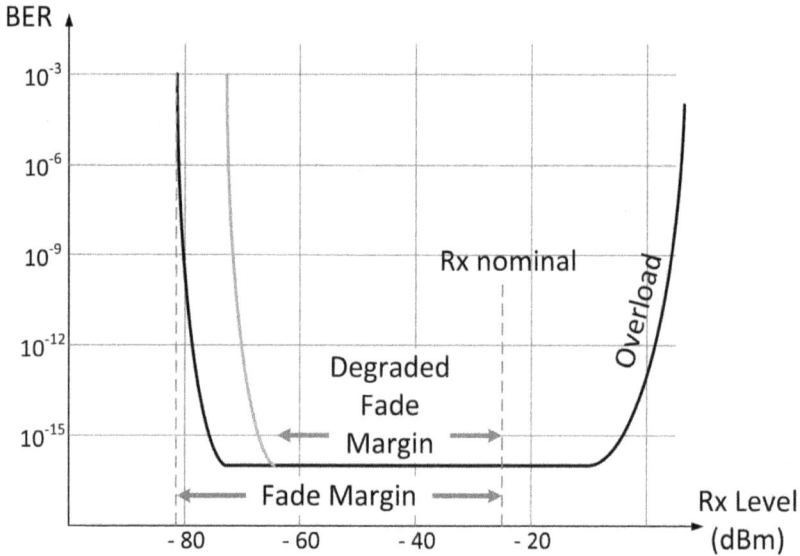

Figure 6.1 Threshold degradation

This aspect will be covered in more detail in the Link Design section, later in the book (Chapter 8). The detailed analysis and set of formulas needed to create interference analysis software are covered in the international telecoms standard ITU-R F.452.

SPECTRUM ALLOCATION

The body that coordinates the international allocation of frequencies, to ensure interference-free operation, is the International Telecommunications Union (ITU), an agency of the United Nations (UN). There are three main departments of interest to radio planners: the ITU-T is the sector that defines end-to-end circuit parameters; the ITU-R is the sector that defines parameters for the radio part of the network, and the ITU-D is responsible for policies, regulation and training programs.

Most standards are available for download via the ITU website www.itu.int/

Frequency plans are created in the ITU-R F.series, and any changes to global frequency allocations are decided at the World Radio Conference (WRC), which is held every 3 to 4 years.

It is up to the government of each country to appoint a frequency regulator who then uses the ITU standards to regulate spectrum use. Two ITU-approved plans may interfere with one another, which can create cross-border complications. It is also the reason that two plans cannot be used simultaneously within an individual country. It is the job of the Regulator to pick one plan, per frequency band, that all operators must use in that country.

Two main types of licencing exist: **_Individual_** licences and **_General Use_** Licences.

With individual licensing, where spectrum is usually controlled and allocated by the country's telecoms Regulator, the specific apparatus is registered and approved for a specific Effective Isotropic Radiated Power (EIRP) output at a specific location. Individual licensing is sometimes referred to as _apparatus_ licensing. In apparatus licencing, each end of a radio link is registered, and detailed interference calculations are done to guarantee the link will be interference free. A fee is paid to the Regulator for this privilege, but in return, if any interference does occur, the Regulator is obliged to resolve such potential interference events.

With general use licensing the equipment apparatus is licence-exempt when deployed, and the regulations for use are covered under regulations for Short Range Devices. The equipment is licenced for the entire class of equipment, but an individual licence is not required each time to deploy. Broad technical parameters of the equipment are defined, but the equipment may be deployed in any geographic location. In the case of license-exempt equipment, the usage of the same spectrum by others is not controlled, leading to an interference risk, and it is the operator's responsibility to resolve it. General or license-exempt licensing is sometimes called *unlicenced*; although I prefer the term *class licence*, as used by some Regulators, as although the entire class of equipment is approved for generic use in any geography, the equipment is still licenced.

A new variant called *light licensing* refers to self co-ordinating bands, such as E-band, where the user defines the channel to be used, and it is registered by the Regulator. There is work being done for a more advanced approach to licensing in future, such as Dynamic Spectrum Allocation and Licensed Shared Access. Under this model, spectrum would be freed up when it was not being used to increase spectrum efficiencies.

FREQUENCY BANDS

Historically, high capacity long-haul links were in bands such as 4 and 6 GHz (often with 11 GHz used for diversity) and 7 and 8 GHz for private operators. Link capacity per frequency channel was just below 200 Mbps. Bands such as 13, 15, 18,

23 and 38 GHz were used for short haul links, with similar link capacity. E-band (70/80 GHz) allows for much wider channels as each band is over 5 GHz wide. For example, 20 Gbps systems are currently being deployed.

Microwave frequency channel spacings in the US were originally based on a 2.5 MHz raster (aggregating to 10/20/40/80 MHz, etc.). In Europe, the raster is in increments of 3.5 MHz (aggregating to 7/14/28/56 MHz/112/224 MHz etc.). A raster of 1.75 MHz is also allowed, as shown in the ETSI EN 302 217 standard, as well as 3.5 MHz itself, and subsets of 1.75 MHz for certain frequency bands.

Today, higher and higher capacities are available through the availability of wider channel plans, for example, 112 and 224 MHz (in 23-42 GHz) and 250 MHz and above (in D and E bands). For example, E-band covers 71 - 76 GHz paired with 81 - 86 GHz (2 x 5 GHz channels), and already using 2 GHz channel plans and 128-QAM modulation 10 Gbps capacity is possible (20 Gbps with CCDP). In D-band spectrum, 5 GHz channels could be made available, thus offering even higher capacities. A mixture of licenced and unlicenced bands can be aggregated to make the configuration more economical to deploy. New single enclosure radio units also allow for CCDP units with XPIC using an Orthomode transducer (OMT) to combine the two opposite polarisation channels in a single polarisation duplexer, thus reducing the physical form factor and cost. In addition to multi-carrier configurations, some manufacturers are also creating multiband integrated antennas to make dual-band configurations more economical to deploy.

FREQUENCY PLANNING

Spectrum is extremely valuable as it is a scarce resource. Operators can pay hundreds of millions of dollars to secure spectrum and with the massive capacity demands from 4G and 5G cellular networks, more and more fixed microwave link bands are being repurposed for mobile use. Spectrum planning is thus essential.

When interference calculations are done, all analysis is assumed to be co-channel. The way adjacent channels are dealt with is to convert them to co-channel via their filter roll-off characteristics called the Net Filter Discrimination (NFD), as shown in Figure 6.2.

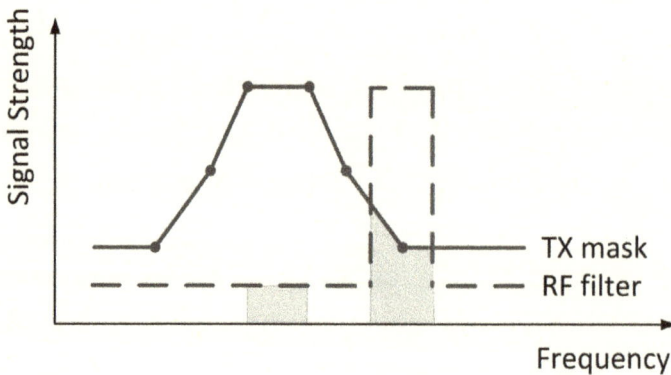

Figure 6.2 Net filter discrimination

Hence:

$$C/I \text{ (co)} = C/I \text{ (adj)} + NFD$$

where

C/I (co)= Carrier to Interference ratio of the co-channel signal

C/I (adj)= Carrier to Interference ratio of the adjacent channel signal

NFD - Net Filter Discrimination

As discussed earlier, in digital systems it is the relative interference level between the faded carrier and the unwanted interferer that is of interest. If the fading event affects the carrier equally to the interferer, then the interference is considered correlated, and the C/I is unaffected by fading, and so the fade margin can be ignored in the calculation. There is some debate as to whether rain fading can be considered correlated or not. Even at a nodal site, it is theoretically possible that the rain cell is on the edge of the node and that the interference path may be unaffected at a time that the main path is rain affected, as shown in Figure 6.3.

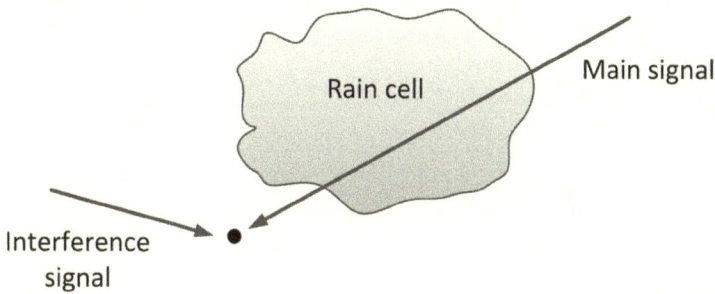

Figure 6.3 Uncorrelated interference

A conservative approach would be to consider that the *wanted* signal could experience rain fading, yet the *unwanted* interferer remains unaffected. Where the carrier fades independently, such as multipath fading, the interference is *uncorrelated*, and the fade margin always needs to be taken into account. In other words, the interference should be analysed at the threshold of the radio.

There are three main interference considerations: bucking interference, nodal interference, and overshoot interference.

Bucking interference, also called *site sense* interference, refers to the interference experienced by a receiver that is operating in the same portion of the frequency band as another transmitter at the local site. The rule is that all transmit signals at the entire site should be tuned to either the high or low end for a particular frequency band. Transmit *low* sites are called A-ends and transmit *high* sites are called B-ends, so site sense planning is also sometimes called A-B planning as illustrated in Figure 6.4.

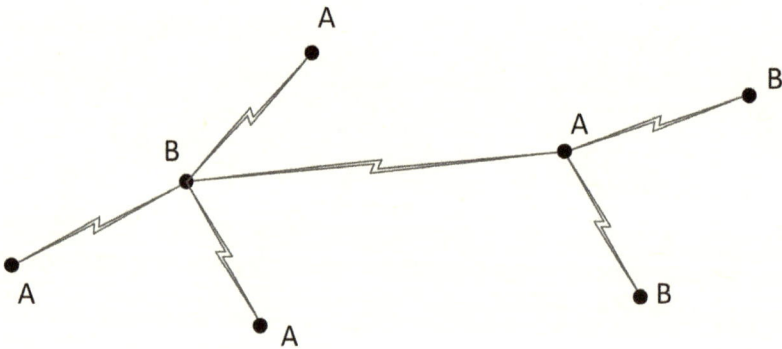

Figure 6.4 Site sense (bucking) planning

All transmitters at an A site transmit low, and so it is not possible for another operator to break through the antenna discrimination and duplexer isolation and cause interference. You may be forgiven for thinking that because you are on a different frequency channel in the band, and the interference is coming out the side of the unwanted transmit antenna into the side of your receive, it will not cause problems. The issue is the dynamic range involved when a transmit signal could be as high as 20-30 dBm, and a faded receive level could be as low as -90 dBm, which is over 100 dB dynamic range! The only way to reduce the transmit signal down to an acceptable level is to use site sense planning and use the filtering inherent in the T-R spacing of the channel plan.

In the past, the frequency Regulator would specify the site sense, so it was not as essential for the operator to check this, but today in many frequency bands such as E-band, it is assumed that the operator is aware of this problem and has planned accordingly. Site sense is also something to be aware of in license-exempt bands as there is no guarantee that there are no high-low site conflicts.

Site sense is also critical with any ring topology as you cannot have an uneven number of links in a loop. Referring to Figure 6.4, it can be seen that the two sites on the right could not be connected with a microwave link in the same frequency band as you would be attempting to connect two B-ends creating a Hi-Lo clash. An additional site would have to be added so that an even number of links were used to complete the ring. Alternatively, a different frequency band could be used, if allowed, where a different site sense could be allocated to the new frequency band.

The second type of interference, **nodal interference**, occurs when we reuse the same frequency channel at a node. The practice of reusing channels as many times as possible before moving to a different channel is good frequency planning. Frequency channels are an expensive and scarce resource and so should be preserved as much as possible. By re-using channels in a planned manner, much greater spectrum efficiencies can be gained. Nodal interference occurs into the antenna of the adjacent link, at the same site, as shown in Figure 6.5.

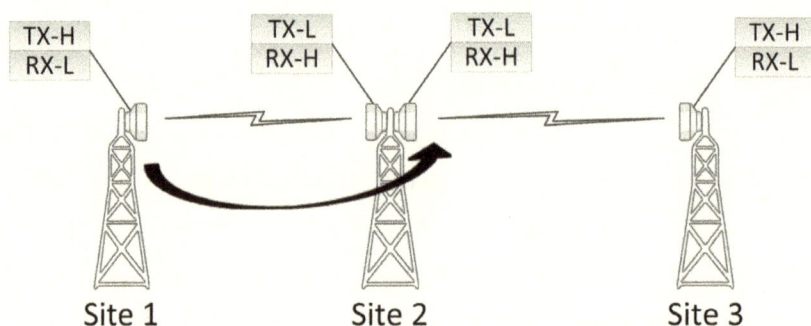

Figure 6.5 Nodal interference

Choosing a high-performance antenna with low sidelobes, and excellent front-back (F/B) ratio allows a planner to reuse a channel multiple times at a node. Bands like E-band with their very small antenna apertures allow a significant number of links to be packed into a node. Referring to an earlier discussion on pencil-thin beams, it is often this insight about the efficiency of frequency reuse at a node that makes planners erroneously think that they can, therefore, reuse the frequency at multiple sites.

In terms of nodal interference, changing polarisation only helps to reduce the interference when the interfering link is directed into the front half of the antenna. The cross-polar discrimination improvement can be determined by referencing the supplier's Radiation Pattern Envelope (RPE) diagrams.

The final type of interference covered here is **overshoot interference**. This type of interference occurs at the subsequent site in a radio route, where the receiver is tuned to the transmitter site two links away, as shown in Figure 6.6. For each site, the transmit high (TX-H), transmit low (TX-L), receive high (RX-H) or receive low (RX-L) are specified.

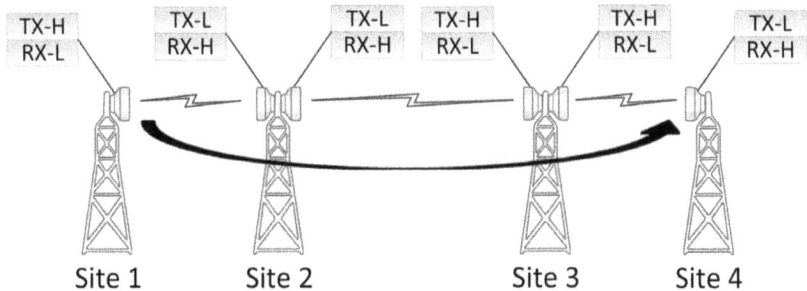

Figure 6.6 Overshoot interference

Despite the interferer being a long way away, distance itself has limited impact on the interference considerations. We have discussed how much discrimination is required between the wanted Carrier to unwanted Interferer signals, and it typically exceeds 40 or 50 dB. Distance helps indirectly through earth bulge, and physical blocking of the line-of-sight, but the interfering signal from a 50 km hop is only reduced by 6 dB, 100 km away, and 12 dB, 200 km away from pure signal degradation. Considering that most long hops over 30 km would have a fade margin in the order of 40 dB and that the minimum Signal to Noise Ratio (SNR) would probably be around 30 dB, a reduction of 70 dB in signal strength would be required before the interference could be ignored. Even at 13,000 km away, the signal would only have reduced by 50 dB, based on distance alone. It is why a microwave signal can happily reach a geostationary satellite 36,000 km away. It is also why more attention should be paid by planners to reduce the antenna height placement and rely on diffraction to protect against interference.

In this case, the interference is into the front of the antenna, and so alternating polarisation is a good strategy.

From the analysis above, it can be seen that an entire radio chain could be designed on a single frequency pair by using the following planning guide:

Channel 1 (H) – Channel 1 (H) – Channel 1 (V) – Channel 1 (V) – etc.

It should be noted that if polarisation alternation is already being exploited for network capacity reasons, for example in a Co-Channel Dual Polarisation (CCDP) mode, then it cannot be used again for interference reduction.

CHAPTER IN A NUTSHELL

Any signal within the passband of the receiver will impact the demodulation process. The signals which were not generated by the microwave radio's transmitter end are considered interference. The basic principle is that the demodulator requires a minimum Signal to Noise Ratio (SNR) to correctly demodulate the signal and so even if an interference signal is so low that it is below the receiver threshold of the radio, it can still increase the noise floor and negatively impact the performance of the demodulation process. In essence, the flat fade margin is degraded so that errors occur earlier than would otherwise have occurred during fading such as rain or multipath ducting. In digital systems, interference is considered a hidden gremlin as the signal level may be so low that it is hard even to measure, and the effect is not seen until fading occurs, which is infrequent.

Frequency bands are allocated globally by a United Nations agency called the International Telecommunications Union (ITU). Each country around the world appoints a frequency regulator who decides which ITU plans to deploy including the local rules of licencing. Two main types of licencing exist: **Individual** licences where the specific equipment and equipment configuration of a specific apparatus is registered and approved at a specific location – also called an *apparatus* licence; **General licences** are sometimes called *unlicenced*, or *class licenced*, as the entire class of equipment is approved for generic use in any geography.

A new variant called *light licensing* refers to self-co-ordinating bands, where the user defines the channel to be used, and registers it with the Regulator. In future, more advanced licensing models such as Dynamic Spectrum Allocation may eventually be used to increase spectrum utilisation. Automated Frequency Coordination (AFC), which is a static sharing solution, is already being considered for the 6 GHz band in the US.

It should also be mentioned that in some frequency bands operators can purchase or reserve a block of frequencies in a specified geographical area. Block licensing is often for Point-to-MultiPoint (PMP) use but includes Point-To-Point (PTP). Examples include 28 GHz, which has been the controversial target of 5G services in some countries. The challenge with block frequencies is that an operator may be allocated spectrum that they do not actually use, which is very wasteful when considered from an objective perspective.

Three main types of interference are of interest: *Bucking interference* which can be minimised by keeping all transmitters and receivers at each radio site either transmitting in the high or low part of the band, but not both. *Nodal interference* which occurs when links on the same frequency transmit into a common node, and *overshoot interference* which occurs into the third site along a link of radio hops. Alternating polarisation to improve interference rejection only works when the interference is into the front of an antenna and so is a proven technique to reduce overshoot interference. It should be noted that if the alternate polarisation is already used for increasing capacity, for example in a CCDP configuration, then polarisation diversity cannot be used for overshoot protection.

7

HARDWARE CONSIDERATIONS

❖ Understand the history of transmission standards (PDH, SDH, ATM and Ethernet)
❖ Understand the benefits of Carrier Ethernet
❖ Understand the basics and trade-offs of digital modulation
❖ Understand equipment features such as adaptive modulation and coding, adaptive bandwidth and adaptive equalisation
❖ Understand the importance of band, carrier and link aggregation and CCDP with XPIC
❖ Know the differences in branching options (HSB, CCDP, SD)

HISTORY OF TRANSMISSION STANDARDS

For those who did not live through the evolution from circuit to packet transmission standards, a historical overview will be discussed to help provide context to the current standards and architecture in microwave radio equipment.

Early transmission networks were all circuit switched, which means that when a connection was established, it was connected until the session terminated, whether any user traffic was being carried or not. In a voice call, during any gaps in the conversation or between words, no-one else could access the channel. Reliability was high, but efficiency was low. The multiplexing methods by radio networks to combine user traffic, such as Plesiochronous Digital Hierarchy (PDH), Synchronous Digital Hierarchy (SDH) and Asynchronous Transfer Mode (ATM) were all based on Time Division Multiplexing (TDM), which uses a hierarchy of dedicated timeslots to combine multiple users onto the aggregate bit stream.

The first digital multiplexing system used on microwave was PDH.

PDH stands for Plesiochronous Digital Hierarchy. *Plesio* means nearly. In other words, it was a system that made the network look like it was synchronous, which is achieved by running the line clock faster than any incoming traffic, and then stuffing dummy traffic in the gaps to ensure the buffers didn't run empty. A coding scheme allowed the equipment to strip off these dummy bits, thus ensuring that all bits were reliably transferred across the network irrespective of their relative individual clock rates.

PDH multiplexers used bit interleaving of the traffic, so one bit from each line stream would be clocked in sequentially. The aggregate stream would then be scrambled to avoid any spectral peaks in the data stream. Transmission was highly reliable, and the timing could be extracted off the aggregate signal making synchronisation easy. In fact, the synchronisation is so good on PDH networks that some operators have kept them in operation purely to carry a reliable clock signal. The downside was that there was no way of identifying the original traffic that was inserted into the multiplexer once it was spread across various bits in the aggregate stream. The considerations discussed above, in addition to the capacity limitation of 140 Mbps, led to the SDH and SONET standards.

SDH stands for Synchronous Digital Hierarchy. The original goal was to have a common standard for E1 (primary rate in Europe) and T1 (primary rate in US/Japan) traffic, but apart from achieving common line rates, and some commonality in the approach, we have ended up with a European-based SDH standard and a US-based SONET standard.

The original idea was to have a fully synchronous standard, where the traffic could be monitored end-to-end and controlled electronically. The way this was achieved was through placing the incoming traffic into a Virtual Container, which could then be directed and reported on anywhere in the network. But this came at a cost! In PDH, 64 E1's could fit in a 140 Mbps aggregate signal, yet in SDH only 63 E1's required 155 Mbps. The 'lost' capacity was all in management overheads.

In addition, it turned out that in most real networks, the synchronisation scheme was unusable. The principle of synchronisation was to create a pointer byte that defined the start of the virtual container. Where any timing changes occurred, the pointer byte could be adjusted to point to a new start point within the frame, thus maintaining synchronisation. The problem was that the pointer byte increments were set at a minimum of 3 bytes (24 bits), so if this scheme were used the network would experience 24 Unit Intervals (UI) of jitter. To avoid this, in most large networks, external synchronisation was required. At a time that network efficiency was required, SDH/SONET was introducing inefficiencies in overhead bytes on top of the circuit switched inefficiencies.

The industry thus looked for a standard that could carry mixed data and voice traffic more efficiently. Quality of Service (QoS) was also desired to enable differentiation of different traffic types, and ATM was developed.

ATM stands for Asynchronous Transfer Mode. It is a connection-oriented, cell switching protocol designed to carry a mix of voice and data traffic. In the context of ATM, a cell is a fixed-size packet of data. In voice traffic, the smaller the packet size, the better the performance due to the real-time nature of voice. In data traffic, the larger the cell size, the more efficient the transmission. To compromise between the two, ATM designers chose a fixed cell size trade-off of 53 bytes which in some ways was not ideal for either traffic type. Importantly, it achieved better efficiencies by using statistical multiplexing to allocate packets to the fixed cell. Being able to manage and report on traffic in a differentiated manner, depending on the traffic type is a significant advantage of ATM, however, as it is still a TDM technology and only quasi-packet based, it was an expensive technology to use. ATM synchronisation is similar to that used in PDH with idle cell stuffing.

The current challenge in the telecoms industry is that the exponential growth in data has placed unprecedented demands on the network capacity. In the past, as capacity demands increased, the revenue associated with that growth grew proportionately. With data, this is no longer the case, as shown in the graph in Figure 7.1.

As an example, to download a video clip on a mobile smartphone may require 1000 times more bandwidth than a voice call, yet the monthly fee for the service may only be fractionally higher than a voice-only plan. The industry turned to a standard developed back in 1973 called Ethernet.

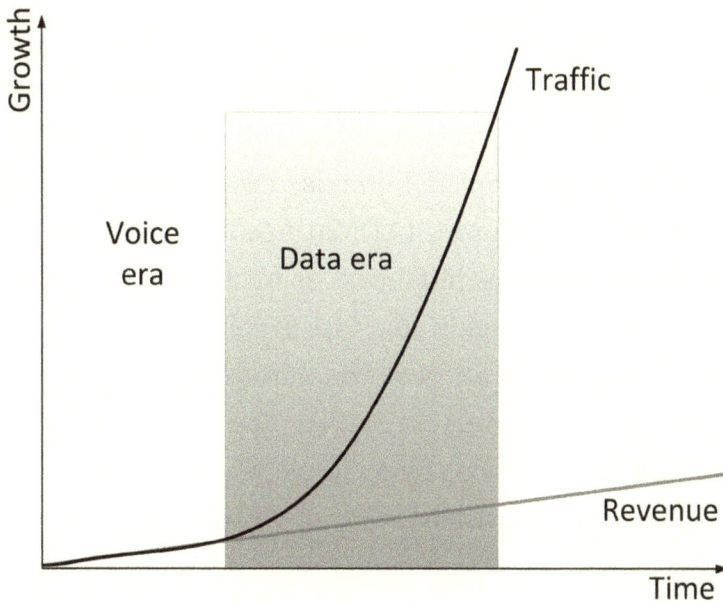

Figure 7.1 Traffic versus revenue gap

Ethernet (IEEE 802.3) is a frame-based data standard that allows multiple PC's and data equipment, such as printers, to share a common physical connection. Compared to the TDM systems already discussed, the transmission efficiencies of Ethernet can meet the exponential traffic burden. The problem was that it was never designed for real-time traffic, it did not accommodate differentiated traffic, with the service-specific quality metrics that ATM provides, nor did it accommodate any service management requirements, in reporting, traffic monitoring or traffic control. An industry organisation called the Metro Ethernet Forum (MEF) thus got to work to create a new WAN standard from the LAN Ethernet standard. The new standard is usually referred to as Carrier Ethernet.

The goal of Carrier Ethernet was to have all the carrier-grade benefits of ATM, yet with the capacity efficiencies of Ethernet. A carrier-grade service needs to be able to support circuits with 99.999 per cent (so-called five nines) reliability and must be capable of switching paths in less than 50 ms, just as was achieved in SDH/SONET. This requirement was specified with the SDH/SONET standard when it was realised that Add Drop Multiplexers (ADMs) needed to restore the circuit in less than 50 ms in an east or westbound protection path, or the end equipment would lose synchronisation.

Carrier Ethernet is an evolving standard, especially in the area of so-called Lifecycle Service Orchestration (LSO). LSO is a set of specifications being developed by the MEF that includes activities like provisioning of the customer service, fault management and end-to-end analytics and reporting of service performance. In practice, services are delivered over the networks of multiple carriers, so the MEF is also specifying the carrier-to-carrier interfaces, and the APIs (Application Programming Interface) required that includes the end-to-end service delivery.

The MEF service quality guarantees include things like availability, frame delays (latency), frame delay variation (jitter) and frame loss (quality and throughput).

Ethernet is an asynchronous technology. Where TDM circuits are carried over frame/packet networks, synchronisation can be an issue. The ITU definitions and terminology for synchronisation in packet networks is defined in ITU-T G.8260.

Various methods can be used to ensure synchronisation of the Ethernet switches including locking them to a Caesium clock, through for example a GPS connection. The ITU also defines an Ethernet synchronisation standard using the physical layer called SyncE. A packet time stamping system called IEEE 1588.v2, which is similar to NTP (Network Time Protocol) used in PC synchronisation, is usually used to allow real-time, synchronous traffic to traverse the asynchronous Ethernet network without errors.

So-called non-line-of-sight equipment is also available on the market today that uses baseband processing to reconstruct the receive signal, thus allowing excellent performance even in fully diffracted or heavy interference environments. Instead of eliminating these rogue RF signals, the rogue signals are captured and used to reconstruct the originally transmitted signal.

Another technology finding its way into microwave link equipment, especially point-to-multipoint systems is Multiple-Input-Multiple-Output MIMO systems. The principle of Multi-user MIMO (MU-MIMO) is to have multiple antenna systems creating multiple paths in so-called *spatial multiplexing*. Massive MIMO is where the number of paths is more than ten times the physical number of end terminals and is currently supported for TDD, although FDD systems are also being developed. Electronic beamforming of the phased arrays used for the antennas can also be combined with steerable antennas as used in 5G technology.

With Software Defined Networks (SDN), which includes SD-WAN, the end-to-end customer experience is less impacted by the performance of individual network standards, because the service is redirected to whatever physical Layer 1 path is available.

BASIC RADIO HARDWARE COMPONENTS

All microwave radio systems are built up of some common building blocks. In its most basic form, there is the Multiplexer-Demultiplexer (MulDem) section, which is the customer interface multiplexing unit, the Modulator-Demodulator (MoDem) section, the Transceiver (Baseband/IF to RF) section, and the Branching unit to connect the equipment to the antenna as illustrated in Figure 7.2.

In the case of All-Indoor radios, only the feeder cable and antenna are mounted outdoors. At the opposite extreme All-Outdoor radios only have a data cable, including the power feed, that terminates indoors, or in a roadside cabinet. Split configuration equipment has the transceiver mounted outdoors in an Outdoor Unit (ODU) with the Muldem and MoDem located in the Indoor Unit (IDU). In some cases, the modem is also located in the ODU. In terms of equipment volumes, all-outdoor and split-mount radio configurations ship significantly more volumes and have the highest growth expectations.

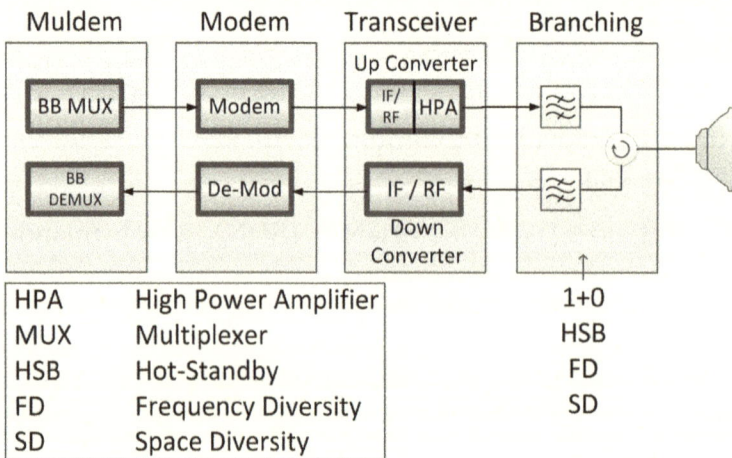

HPA	High Power Amplifier	1+0
MUX	Multiplexer	HSB
HSB	Hot-Standby	FD
FD	Frequency Diversity	SD
SD	Space Diversity	

Figure 7.2 Hardware building blocks

DIGITAL MODULATION

The basic concept of transmitting signals over an RF carrier is somehow to encode the high-frequency radio carrier signal with a baseband signal that can be extracted at the other end of the transmission. Bandwidth efficiencies are also sought after to ensure that the maximum amount of data can be transmitted using the lowest amount of bandwidth.

There are various forms of digital modulation. Digital modulation is achieved by adjusting the analogue RF carrier to represent the digital baseband data that is superimposed on it. Adjustments include Frequency Shift Keying (FSK) where the carrier is adjusted by positive and negative frequency increments to represent the 0's and 1's in the data stream. These shifts in frequency can be analysed at the opposite end to decode the data being transmitted. A similar principle is used in Phase Shift Keying (PSK), where a shift in phase is used to code the binary data, as shown in Figure 7.3. Most microwave radio systems use Quadrature Amplitude Modulation (QAM), which is a mixture of both Phase and Amplitude Modulation, as shown in Figure 7.4. Multiple bits are grouped in a unique constellation to reduce RF bandwidth requirements. As the number of bits per symbol is increased, the RF efficiency goes up but at the expense of quality as illustrated in Table 7.1.

In the BPSK constellation represented in Figure 7.3, it would require +/- 180 degrees of jitter to decode the bits incorrectly. In the QPSK system, only +/- 45 degrees can be tolerated.

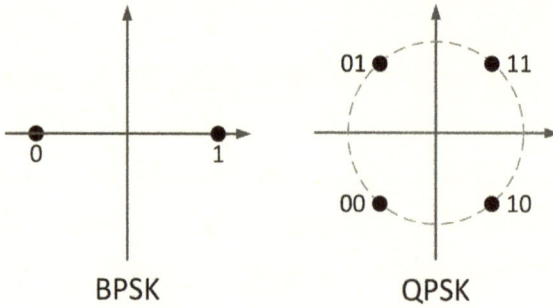

Figure 7.3 Digital modulation constellation

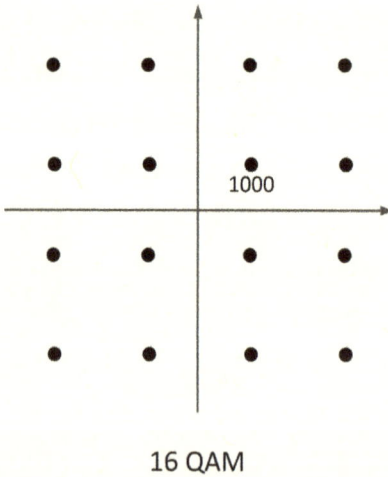

16 QAM

Figure 7.4 Quadrature Amplitude Modulation

For illustration purposes, Table 7.1 compares the multi-QAM systems for a 100 Mbps signal, to illustrate the bandwidth vs quality trade-off.

It can be seen from Table 7.1, that initially as QAM level is increased, there is a significant bandwidth reduction, but there are diminishing returns for the very high-QAM schemes. Despite this, manufacturers have 4k-QAM systems available and are already planning 8k-QAM and higher. What is often forgotten is that for each additional bit added to the symbol, there is approximately a 3 dB reduction in receiver sensitivity (halving of system gain), in addition to a further reduction from transmitter backoff to maintain linearity in high-QAM mode. Considering that the minimum SNR requirements increase significantly with high-QAM modes, in practice the very-high-QAM operation may be not able to be used due to the quality degradation, given the limited bandwidth savings achieved. One exception may be for very short hops.

Table 7.1

Bandwidth reduction limitations

Modulation	Required BW	Bits/s/Hz	System Gain
4 PSK	50 MHz	2	Reduces > 3 dB
16-QAM	25 MHz	4	Reduces > 9 dB
64-QAM	17 MHz	6	Reduces > 15 dB
128-QAM	14 MHz	7	Reduces > 18 dB
2048-QAM	9 MHz	11	Reduces > 30 dB

ADAPTIVE CODING AND MODULATION (ACM)

In the past, microwave radio link designers were faced with the tradeoff between capacity and quality. Once they had made their choice, they had to live with that choice for the lifetime of the link. Adaptive modulation is a scheme whereby the link modulation scheme can be automatically changed to adapt to any weather or interference, as shown in Figure 7.5.

With adaptive modulation, designers are now able to get the best of both worlds. In Figure 7.6, the impact on various radio links is shown, for a few fades with different fade depths.

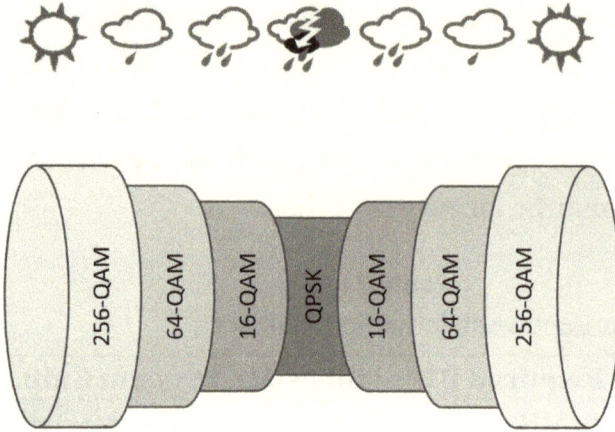

Figure 7.5 Adaptive coding and modulation

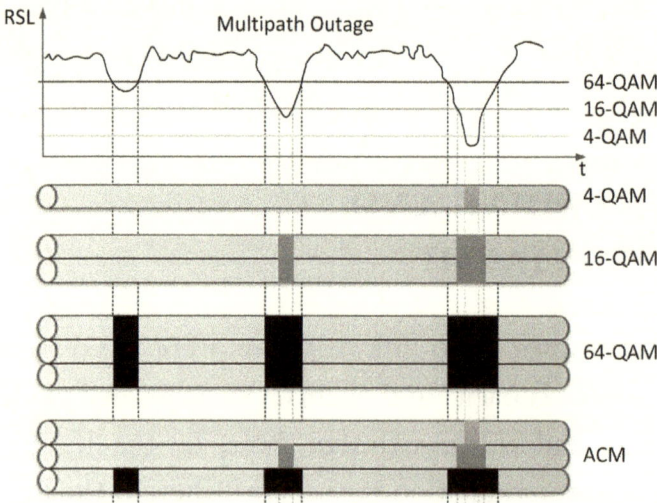

Figure 7.6 Fading impact on error performance

The lower part of Figure 7.6 compares fixed modulation schemes to an ACM scheme. It can be seen that triple the capacity of a 4-PSK link can be achieved by moving to 64-QAM, however, with more outages. Considering the supervisory signals, as well as link synchronisation, are linked to these channels, the longer outage shown in the 64-QAM example may not meet carrier grade standards, and therefore a lower capacity option would be chosen. With the ACM example, all priority traffic, plus the supervisory and synchronisation signals can be linked to the top 'virtual pipe' thus achieving similar performance to a 4-PSK system, yet the rest of the time the additional capacity can be enjoyed, albeit at a lower performance level. Bear in mind that in the 4-PSK example, the full RF bandwidth is occupied for 100 per cent of the time, with that lower capacity. The signal outages experienced on the lower two 'virtual pipes' are on traffic that was not available at all in the 4-PSK example.

Another use-case for ACM is to extend the link for as long as possible during deep fades. Some operators plan the throughput at the required availability and use that as the reference modulation for the calculations.

A common misunderstanding based on the name, *adaptive* modulation, is to assume that somehow the link is being constantly adapted to changing weather conditions. Network planners are thus unnecessarily concerned that the transmission medium is not stable, which complicates switching mechanisms at higher layers of the network stack. In reality, the changes only occur on the rare occasions that, by not switching, an outage would have occurred. In carrier grade systems this is less than 1 per cent of the time.

Apart from any price premium that the radio manufacturer may add for licencing, an ACM option only has an upside. The way to design these systems is to imagine that the link consists of parallel links, each on a different modulation scheme, and then design each 'link' accordingly.

In addition to adaptive modulation, modern chipsets incorporate *adaptive coding and adaptive bandwidth*. Adaptive coding allows the coding scheme to be set by software to capitalise on the tradeoff between capacity and quality. A system with very low coding will have the lowest latency and the highest capacity, but the lowest performance. A system with high coding will sacrifice some latency and throughput to achieve increased performance. By using adaptive coding, the appropriate coding scheme can be matched to the desired application.

Some vendors apply adaptive bandwidth and coding to maintain link integrity for as long as possible. Adaptive bandwidth can be applied once the modulation and coding scheme has been reduced to its minimum level. This lower modulation scheme together with a significantly reduced RF bandwidth and hence concentrated RF power can keep the link running through a deep fading period, albeit at a lower throughput. The availability of self-coordinated bands such as E-band, allows advanced equipment enhancements such as adaptive bandwidth to be readily implemented, thus significantly increasing the availability of these links during adverse weather conditions.

AUTOMATIC TRANSMIT
POWER CONTROL (ATPC)

The design methodology in microwave radio links is to run the links with a higher transmit signal than needed, so that during fading conditions the reduction in signal level does not result in an outage. For this reason, for most of the time, the radio is transmitting more power than it needs to, leading to unnecessary interference into other links. In some equipment, this results in the transmitter circuitry running hotter than necessary, with the associated reduction in Mean Time Between Failure (MTBF) of the circuitry.

Automatic Transmit Power Control (ATPC) is a system whereby the transmit signal is backed off through software, during normal operating conditions and only boosted back to full power during fading events. Remote Transmit Power Control (RTPC) is usually implemented through a software setting in the radio equipment. The power level of the remote transmitter can thus be set via software.

In the US, the regulator allows the power to exceed the maximum EIRP levels on the licence, for a limited period. In Europe, the power needs to stay within the rated maximum EIRP on the licence, under all conditions.

AGGREGATION

The word aggregation is used in the industry in multiple ways with very different meanings based on the context. In some cases aggregation refers to a network segment, in others it means a nodal site where the traffic from multiple links are optimised into a combined signal, in others, it refers to combining links of different frequency, polarisation or even technologies.

Ethernet link aggregation is useful where the individual link capacity of the radio channels is less than the aggregate capacity of an Ethernet switch port as defined in IEEE 802.1AX (formerly 802.3ad). Layer 2 aggregation uses MAC addresses to assign traffic.

The concept of Ethernet link aggregation is illustrated in Figure 7.7.

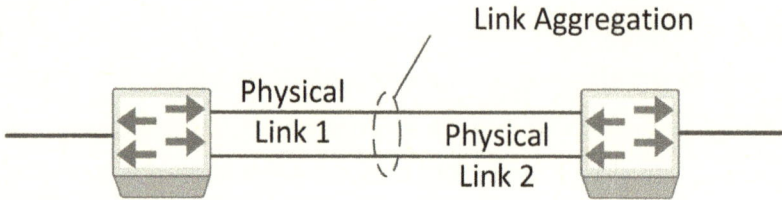

Figure 7.7 Link aggregation

When combining channels of similar bandwidth, the *throughput* - peak as well as aggregated - is very important and so a Layer 1-based method is supported. Layer 1 aggregation often includes load balancing of the radio channels. When combining channels of different bandwidth, such as E-band with 15, 18 or 23 GHZ, a Layer 2/3 based method is used.

The demand for more and more capacity will drive the standards for multiple bands and multiple carrier aggregation - called BCA - as discussed in Chapter 4 (Fading Effects). Various use cases exist for BCA. Historically combining 4,6 or 7 GHz channels with 11 GHz is a well-known technique. Where channels are available and affordable, various mmWave bands can be combined in a 2+0 configuration, although what is even more beneficial is to use lower mmWave frequency bands to back up the very high capacities available in E-band. Considering that many Regulators have distance limitations on certain frequency bands that would not allow BCA combinations to be used, a popular option for BCA is to combine a licensed channel with an unlicensed one. Some manufacturers are providing dual-frequency horn feeds into a single antenna reflector, for example, E-band with 15, 18 or 23 GHz. Dual-band antenna operation can significantly reduce costs, due to hardware cost reductions, as well as the reduction of co-location costs. Additional savings can be made through the reduction in planning complications, such as tower loading, on a radio tower. Dual frequency active antennas are also being developed.

Link aggregation standards are linked to SDN to enable a fully programmable microwave transport network optimised for carrying different traffic types. Equipment is available on the market that aggregates both licenced and unlicenced bands of equipment to exploit the non-correlated nature of fading effects such as rain or atmospheric ducting. Solutions are also developing to be able to use non-contiguous channels to optimise capacity versus the available RF spectrum. Many of these solutions, to create one giant virtual pipe of capacity for x-hauling 5G traffic, will become available on the market for any application or industry.

CROSS-POLAR INTERFERENCE CANCELLER (XPIC)

When fading occurs on a radio link, it can cause phase rotation of the signal. Where dual-polar operation is used, such as CCDP (Co-Channel-Dual-Polar), this can result in cross-polar interference between the traffic channels during fading, thus reducing the performance of the link. To avoid this, manufacturers use a Cross-Polar Interference Canceller (XPIC) that dynamically keeps the two vectors orthogonal to one another, thus reducing the interference risk.

ADAPTIVE EQUALISATION

Transverse Adaptive Equalisation (TAE) is often incorporated on wideband radios, where selective fading is expected. Most

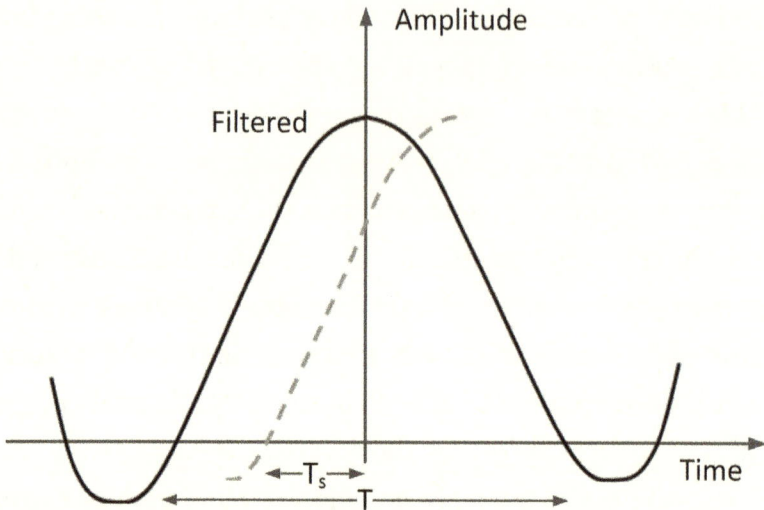

Figure 7.8 Nyquist filter characteristic

designers of microwave radio equipment use a Nyquist filter to bandwidth limit the baseband signal without causing errors. The filter characteristic is such that the leading and lagging tails of the adjacent pulses are at zero at the time of demodulation, as illustrated in Figure 7.8. During multipath fading, the signal becomes distorted, and so instead of these pulses being at zero, they interfere with the symbol being demodulated. The principle of adaptive equalisation is to delay both the leading and lagging edges dynamically so that they are at zero at the time of demodulation.

In Figure 7.8, the lagging tail from the pulse that has just been demodulated, shown as a dotted line, can be seen interfering with the pulse (or symbol) that is currently being demodulated. If this lagging tail were delayed by time, Ts, then it would be at zero when the current pulse is demodulated, thus eliminating the InterSymbol Interference (ISI).

The more delay lines that are incorporated in the equipment, the better the performance is against selective fading from multipath. The so-called Signature curve, shown in Figure 7.9, is used to quantify the improvement in performance.

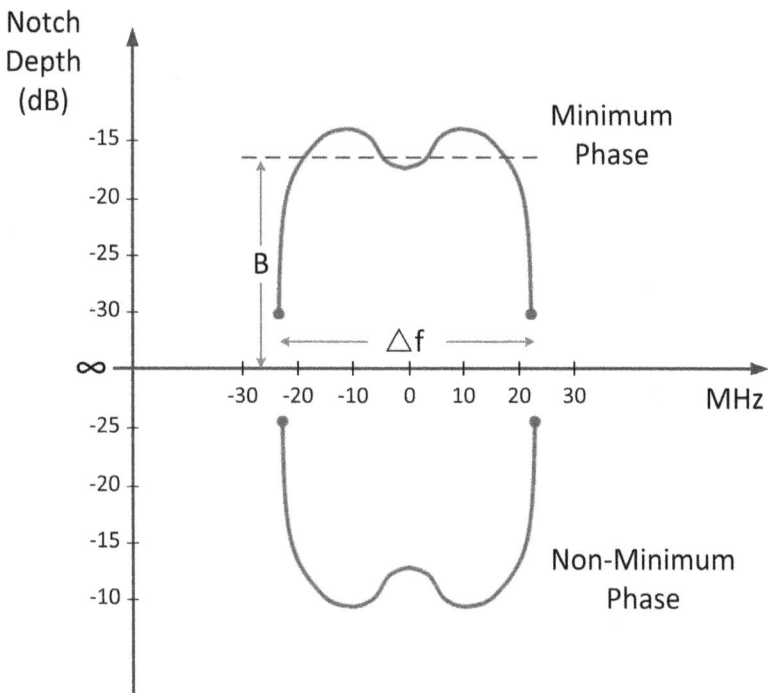

Figure 7.9 Signature curve

The way the signature curve is produced is based on the Rummler model. A signal is injected into the radio receiver at the edge of the band and then swept in frequency across the entire bandwidth of the receiver (Δf). The same signal is delayed by a fixed time-period of 6.3 ns and increased in amplitude (b) until an outage occurs.

$$\text{Notch depth B} = -20 \log (1\text{-}b) \text{ dB}$$

where b is the amplitude of the delayed signal relative to unity.

By averaging the area under the curves for both positive as well as negative B values, a signature value is established. Details of the signature curve can usually be obtained from manufacturers on request in the form of a Dispersive Fade Margin (DFM). The higher the value, the better the performance against selective fading.

BRANCHING OPTIONS

The transmitters and receivers in a radio system are connected to a common antenna through a unit that is known as a duplexer. Various arrangements for long-haul radio systems are shown in the section below.

The simplest coupling mechanism is for Non-Protected (NP) radios, which is also called a 1+0 configuration, as shown in Figure 7.10. The duplexer performs an isolating as well as coupling function. It isolates the transmit signal from its *own* receiver, and couples the *wanted receive signal* from the opposite end into the same antenna that it is using for the *wanted transmit signal* at the local end.

The limitation of a 1+0 (non-protected) configuration is that any hardware failure results in an extended outage. To reduce the outage time, some standby hardware can be made available during failure conditions in a configuration known as

Hot Standby (HSB). The second transmitter is active, hence the term hot, but as the system operates on a single frequency pair, it transmits into a dummy antenna load. Both receivers are active, and a switch selects the working receive signal. The configuration for a Hot-Standby (HSB) system is shown in Figure 7.11.

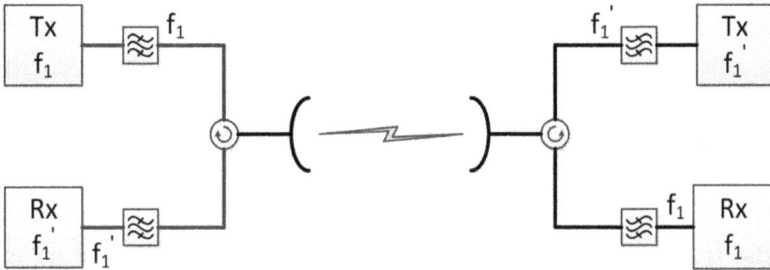

Figure 7.10 Non-protected configuration

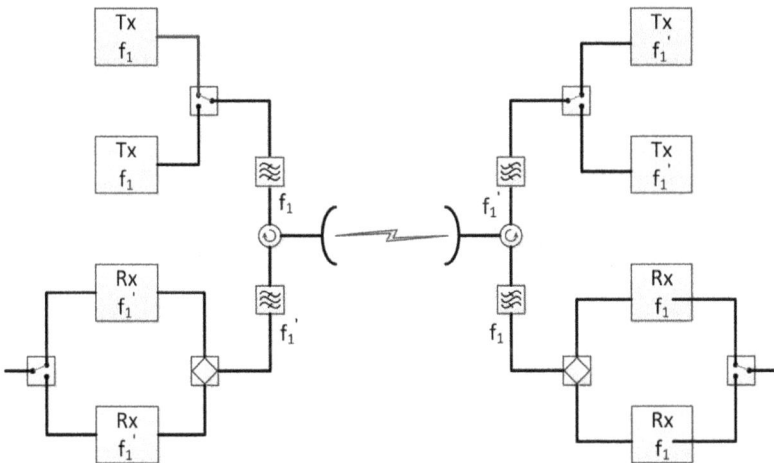

Figure 7.11 Hot-standby configuration

A disadvantage of HSB operation is the loss that is experienced in the receiver branching. If the power is split evenly between each receiver, 3-4 dB of loss occurs, which is equivalent to more than a 50 per cent reduction in antenna size. Some systems use a directional coupler instead of a hybrid splitter to reduce the losses in the primary path, which is a much-preferred configuration considering how seldom the standby system is used. In modern equipment, it is not unusual for the field MTBF to be 30 years or more.

A popular configuration to use is CCDP which uses both polarisations of a channel, which are combined to double the capacity of the link or act as a natural standby path. Because there are effectively two RF channels available, each operating on the opposite polarisation, combining rather than switching is deployed and the associated branching losses are thus significantly lower. Even though the alternate polarisation is effectively an extra RF channel, it is not as valuable as an adjacent channel due to cross-polar interference. For this reason, in many cases, the cross-polar channel is either discounted or even allocated free-of-charge, by some Regulators.

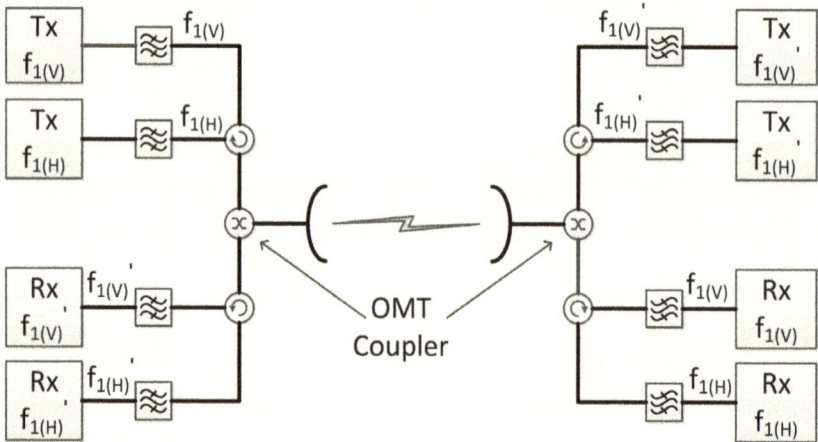

Figure 7.12 Co-Channel Dual Polar configuration

The configuration for a CCDP system is shown in Figure 7.12.

CCDP is effectively a Frequency Diversity (FD) configuration but without the spectrum wastage. The frequency wastage of 1+1 FD is why frequency diversity is not allowed in most countries. What is permitted is N+1 frequency diversity, where the protection channel is shared among multiple carriers, as is shown in Figure 7.13.

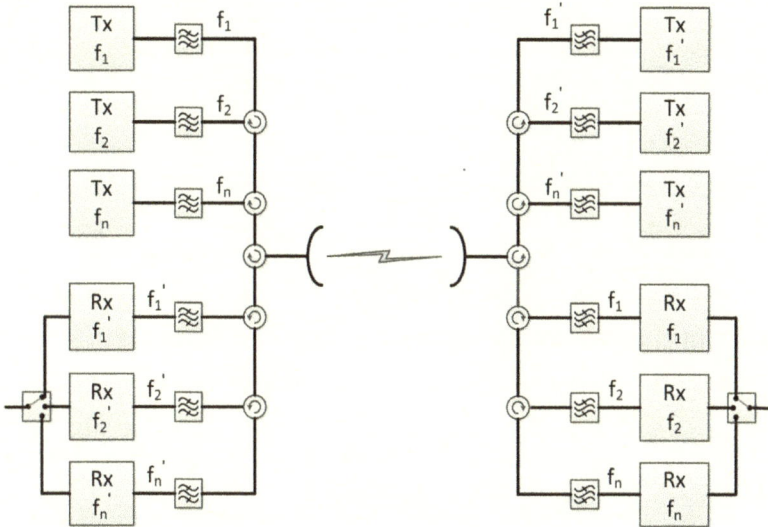

Figure 7.13 Frequency diversity configuration

Space diversity is another branching arrangement that is used as protection against multipath fading as well as specular reflections. The principle of operation is to have two receive antennas that have a different path length to the reflection point, thus ensuring that the interference pattern is not correlated, in which case combining or switching between the two signals eliminates the effect of the stray signal as is shown in Figure 7.14. One manufacturer has deployed an innovative approach where beam-forming and baseband processing is used to eliminate one of the antennas thus achieving full space diversity in both directions using only three antennas. It should be noted that this is different from the hybrid-diversity, 3-antenna system, that has been around for decades. In hybrid diversity both space and frequency diversity are used to eliminate the fourth antenna.

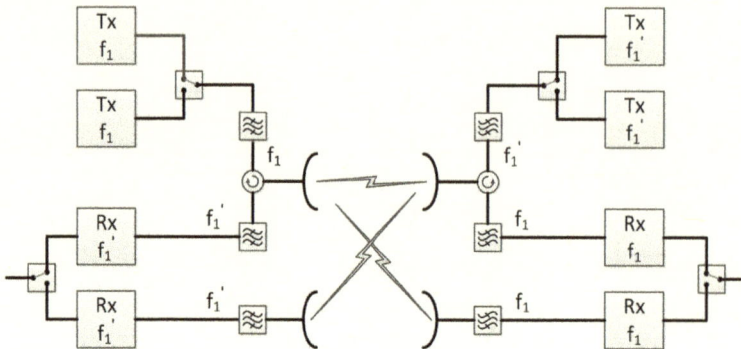

Figure 7.14 Space diversity configuration

CHAPTER IN A NUTSHELL

Microwave hardware has evolved into the current packet-based equipment. Early PDH systems were highly reliable and carried the synchronisation clock in the aggregate bit stream, but they lacked traffic manageability. The SDH/SONET and ATM systems that replaced PDH were still based on Time Division Multiplexing (TDM) and lacked the efficiencies of packet-based systems. The 'smarts' of TDM were incorporated in a new Ethernet standard called Carrier Ethernet, which was developed by an industry body called the Metro Ethernet Forum (MEF).

All microwave systems in principle consist of some common architectural building blocks. The Muldem section is the interfacing and combining unit of various forms of traffic into and out of the radio. This baseband signal is fed to the Modem section where modulation and demodulation onto an RF carrier occurs. Most microwave systems use multi-level QAM, and there is a trade-off between quality and bandwidth efficiency as the modulation levels increase. In practice, there is a diminishing bandwidth efficiency improvement as the modulation levels are increased, so very high-level QAM systems are only really beneficial for very short hops. Adaptive Coding and Modulation (ACM) allows the capacity benefits of multi-QAM systems to be realised without the quality reduction, while adaptive equalisation has virtually eliminated the problems initially experienced on wideband radio systems due to selective fading. The measure of adaptive equalisation effectiveness is measured by a signature curve which is characterised by a DFM value, which can be obtained on request from the manufacturer for each modulation scheme used.

Various forms of branching allow different configurations to be deployed, including Co-channel Dual Polarisation (CCDP) operation and space diversity, significantly improving performance and capacity.

8

LINK DESIGN

- ❖ Know the ITU and industry design standards and objectives
- ❖ Know the difference between Availability and Performance
- ❖ Know what a hypothetical reference path is
- ❖ Understand the variables and importance of P0
- ❖ Know when rain fading and multipath fading will be dominant
- ❖ Be able to do a basic link design, understanding all the main variables used
- ❖ Do a path budget to determine the expected unfaded receive level
- ❖ Predict outage from fading (rain and multipath)
- ❖ Know which countermeasures and configurations to use for different types of outages

INTRODUCTION

Due to the relatively poor performance of non-carrier grade wireless systems on the market, as well as poorly designed carrier-grade systems, there is often a perception that a cable system is more reliable than radio. If your home Wi-Fi network is going up and down, there can be a strong desire to put in an Ethernet cable and get a predictable and solid connection.

Wireless in itself is not the problem. We reliably speak to each other over the air, and planes rely on wireless technology to land safely. In some ways, a non-cable system is more reliable. With a cable break, the system goes down entirely. A rain fade may cut the microwave link for some seconds or even minutes, but then restore itself as rain intensity falls.

These days, to design a radio system to carry traffic error-free is trivial. Digital radio technology is incredibly robust, and interference is usually only an issue in a faded condition, with little or no effect under normal operation. Perversely it is this robustness that has created a performance issue in installed systems. In analogue networks, any error in design or installation was evident on the day of commissioning. In digital systems, the link may be inadequately designed, the connecting cable could be damaged, the antenna could be misaligned, and some serious interference problems could be present, yet the link would run error-free, and so there may be no attempt to fix these issues. These problems would only become evident during poor weather, resulting in reduced operational performance. During the poor performance periods, visible things like rain, dust, or wind may be unfairly blamed as the primary cause. A well designed and installed system would not be unduly affected by these external conditions. This chapter addresses how to design a reliable microwave system to withstand all atmospheric and weather conditions.

RELIABILITY STANDARDS

The first challenge is to work out to what standard to design the link. Historically the ITU provided end-to-end standards that could be used to design an individual radio hop. The ITU-T, which covers international end-end standards that can be used for all the various technologies that the traffic may travel over, defined a 27,500 km hypothetical reference connection for an international circuit (ITU-T G.827) as shown in Figure 8.1. The concept was to provide a reference model for a typical international telephone call. A call could be initiated at a telephone in one country, travel through the local twisted copper connection to the local exchange and then routed in a national trunk circuit through various exchanges. The connections could be microwave radio or fibre. The national link would end up at an international gateway that may be connected to an undersea cable or satellite system, and then the process would be reversed at the country which was terminating the call.

The ITU-R assumed that 2,500 km of this 27,500 km end-end circuit would be over microwave radio, and thus apportioned their objectives as a subset of this. They further divided the connection up into long-haul, short-haul and access as shown in Figure 8.2.

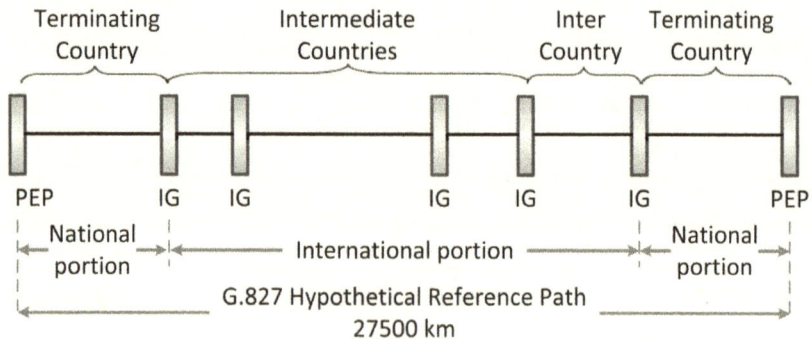

Figure 8.1 ITU-T hypothetical reference path

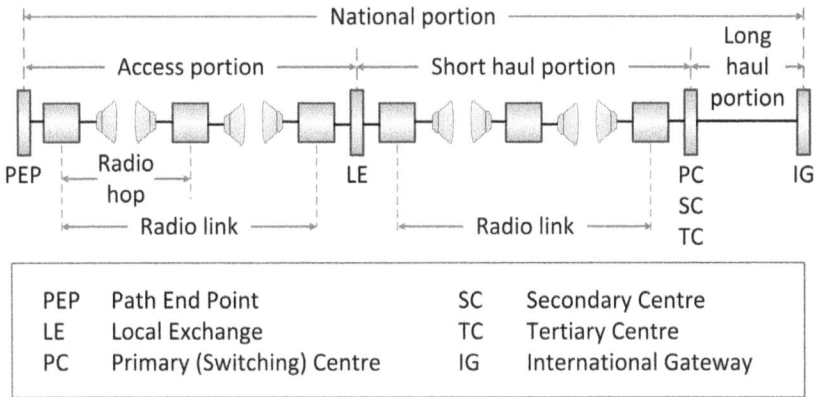

Figure 8.2 ITU-R reference connection

When designing a radio hop for a traditional fixed carrier, the radio planner could use the scaled-down objective as the target and design the hop accordingly.

In today's networks, the architecture is not as simple or predictable, but the logic behind how the hypothetical reference paths are defined helps to set meaningful objectives in current, real networks.

LINK RELIABILITY

Before the 1980's, only one metric was used to define reliability, and that was the average Bit Error Rate (BER). The problem with this metric is that if a serious break in transmission occurred, the millions of errored bits accumulated during the outage would significantly degrade the average performance. A service provider who was achieving a quality level of, for example, 10^{-12} for weeks on end could drop down to 10^{-9} due to one serious outage. In reality, during that outage period, the service would be rerouted, and so it was unfair to say the average quality was 10^{-9}. For most of the time the quality was 10^{-12}, and for a short period, the service was not usable.

For this reason, the ITU defined two metrics: The first one, called **performance,** related to the average quality of the connection when it was usable; the second one, called **availability,** related to the period the service was considered usable. A circuit was only regarded as usable ten consecutive seconds of severe quality impairment, usually measured by an error rate worse than 10^{-3}.

The practical reality of these new standards is that if a service is unusable for nine seconds but becomes usable again for one second, the service is considered 100 per cent *available,* during those 10 seconds. While the customer would be told their service is 100 per cent available, the poor quality experience would be reflected through the *performance* standards.

The quality definition of the ITU standards also means that during an unusable period, the *performance* reporting is 100 per cent, as performance is only defined during available periods. The lousy customer experience, in this case, is dealt with under the *availability standards.*

Table 8.1
Outage type classification

Availability	Performance
Rain, ducting and diffraction	Flat and selective fading from Multipath
Equipment MTBF and MTTR	Equipment dribble errors (BBER)
Power failure	Interference
Catastrophe (fire or tower falling)	Wind

The types of outages in radio systems, affect each metric differently as shown in Table 8.1. The criteria to determine whether the outage affects *availability* or *performance* is whether the fading events last more or less than ten consecutive seconds.

Regarding frequency bands, rain - a slow fading event lasting longer than ten seconds - mainly impacts links above 10 GHz, and hence *availability* is the key metric. Long links generally only occur in bands below about 15 GHz, and thus multipath impacts *performance* of those links as illustrated in Figure 8.3.

Strictly speaking, to design a link to ITU standards, these two separate design parameters should be met. *Availability* standards for the microwave radio network are defined in ITU-R F.1703, last updated in 2005, and are based on ITU-T G.827. The *performance* standards for microwave are defined in ITU-R F.1668, last updated in 2007, based on ITU-T G.828.

Various industry bodies such as the MEF, ETSI and the ITU have been working on standards to create new packet-based standards. The latest ITU standard is ITU-R F.2113, created in 2018, which attempts to map Bit Error Rates (BER) of non-packet links (PDH, SDH), with packet links (Ethernet), where a Frame Error Rate (FER) is relevant. Due to differing packet sizes, this can all get very theoretical, and complex, very quickly and so in practice the traditional non-packet calculations methods are usually used. Based on some theoretical calculations done by

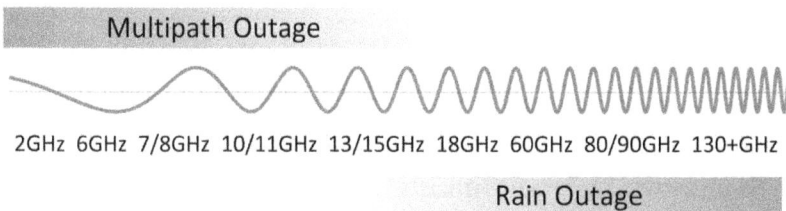

Multipath Outage

2GHz 6GHz 7/8GHz 10/11GHz 13/15GHz 18GHz 60GHz 80/90GHz 130+GHz

Rain Outage

Figure 8.3 Fading type as a function of frequency

ETSI, it may be prudent to increase the fade margin by up to 3 dB for packet-based radio links.

In practise, even in non-packet networks the industry ended up defining a new parameter, that is not defined in the ITU, called all-cause annual reliability. To achieve this, the one-way, worst month performance metric, was modified to a two-way, annual figure so it could be added to availability. A non-standards-based, albeit practical target, of somewhere between 99.99 per cent and 99.999 per cent (with 99.995 per cent being the most common) is typically used as the per hop target.

It is essential to stress that this number is not directly linked to the sacrosanct five-nines network availability goal of telecoms networks. Understandably the two get confused because they appear to be the same number, but it is critically important to understand the difference between the two.

Five-nines network reliability refers to the annual uptime of the network. The reliability number that most radio-planning software programs report is the sum of the availability outages due to rain or extended multipath events, as well as the accumulation of short sub-ten second outages from multipath fading, on an individual hop. No account is made for any redundancy, or path protection schemes, and it is *redundancy* that allows telecom networks to achieve carrier-grade performance or so-called *five-nines* reliability.

The *monthly* circuit availability also tends to be within an order of magnitude of 99.999 per cent, yet there is no direct correlation between the radio outage quoted in radio planning software, and network or circuit performance.

If an individual radio link designed to less than 99.999 per cent was coupled with another link, also designed to less than 99.999 per cent, by combining the two, a circuit availability of 99.999 per cent, or greater, could be achieved. When deploying Adaptive Coding and Modulation (ACM), this is essential to understand. Otherwise, the benefits may not be realised. Fibre networks can only achieve 99.999 per cent uptime if they have circuit redundancy built in, as a single fibre cut would exceed the annual target. Radio networks should be judged by the same yardstick.

Due to the complexity of applying the standards discussed, for practical purposes, most operators set the hop reliability goal somewhere between 99.99 per cent and 99.999 per cent to do the initial design. Before changing antenna sizes, equipment configurations or antenna heights, the network topology should be considered to balance the cost of these enhancements versus network outage risks.

LINK DESIGN

A radio link design needs to be done to meet the reliability standard set in the previous section. In this next section, the various design elements will be covered. The steps to do the design are summarised below, which will be covered in more detail in this chapter:

- Start with an accurate path profile
- Program the radio parameters such as transmit power, receiver threshold, Dispersive Fade Margin (DFM) etc. into the planning software
- Calculate the nominal receive signal level and Flat Fade Margin (FFM) *Note: In some countries, the FFM is set by the Regulator.*

- For short hops in higher frequency bands, ensure FFM meets reliability target for the rain region of the link
- For longer hops in lower frequency bands, increase flat fade margin by increasing antenna size or configuration or increase space diversity antenna spacing, until the link reliability meets the link performance objective, based on predicted Po value
- Choose equipment with the right Dispersive Fade Margin (DFM) to meet the dispersive (selective fade) outage
- Ensure there is limited receiver threshold degradation from interference, and calculate the Interference Fade Margin (IFM) that will impact the overall Effective Fade Margin (EFM)
- For Adaptive Coding and Modulation (ACM) links, calculate each modulation scheme as though it were a parallel link
- Antenna heights should be initially checked using ITU clearance rules but then placed at a practical height that provides the best overall performance balancing multipath and diffraction risks.

The link design formulas required to design a radio system are defined in ITU-R P.530. The main parameters discussed are summarised below:

Propagation effects to take into account:

- Diffraction fading
- Attenuation from atmospheric gases
- Multipath fading
- Surface reflection fading

- Attenuation from precipitation and solid particles
- Angle of arrival fading
- Cross-Polar discrimination deterioration from multipath
- Signal distortion (selective fading) during multipath
- Upfading
- Scintillation fading

FADE MARGIN

The basic methodology is based on achieving a nominal Receive Signal Level (RSL) that is sufficiently high above the radio receiver threshold so that any impairments that occur outside the error-free zone of the radio system cause outages only within the limits of the annual path objectives. This range of RSL in which error-free performance can be achieved is called the Flat Fade Margin (FFM), as shown in Figure 8.4.

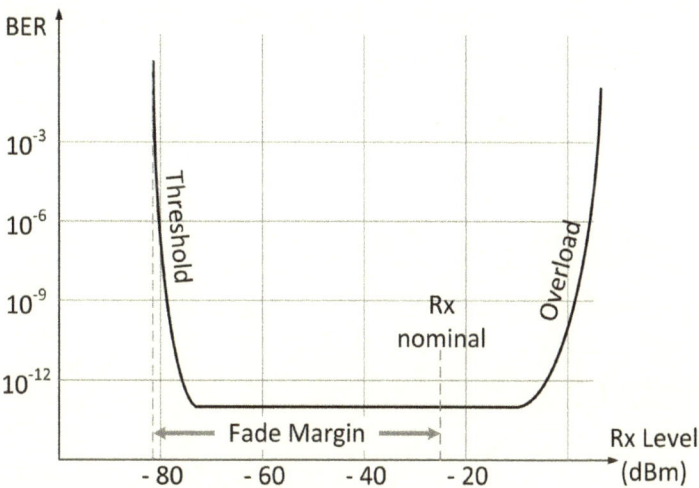

Figure 8.4 Flat fade margin

Flat Fade Margin (FFM) = Rx nominal - Rx threshold

The overall performance is also affected by interference and dispersive fading; hence, an Effective Fade Margin (EFM) needs to be calculated.

Effective Fade Margin (EFM) = FFM +DFM -IFM

where IFM is the Interference Fade Margin caused by threshold degradation. In practice, in normal operation with frequency reuse, the IFM should not exceed 2 dB. The units for Fade Margins are dB.

RECEIVER THRESHOLD

The receiver threshold of the radio, or receiver sensitivity, is the RSL at which the radio effectively becomes unusable. The threshold is usually defined at a BER (Bit Error Rate) of 10^{-3} or 10^{-6}. The value is specified on the microwave radio manufacturer's data sheets and is calculated using the formula below, as illustrated in Figure 8.5.

Receiver Threshold [dBm] = SNR_{min} + NF + kTB

where

SNR_{min} = Minimum demodulator signal to noise ratio
NF = Receiver front-end noise figure
kTB = background thermal noise
k = Boltzmann's constant (1.38×10^{-23} J/K)
T = Receiver temperature in Kelvin (290 K=17° C=room temp)
B = Bandwidth of the receiver

The formula can be simplified, using the data above, as follows:

$$\text{Receiver Threshold [dBm]} = \text{SNR}_{min} + \text{NF}$$
$$\text{[dB]} - 114\ \text{[dBm]} + 10\log\ (\text{B [MHz]})$$

The radio receiver threshold is often misunderstood in the industry. Understanding the history of quality standards puts some perspective on the subject. When voice was the primary traffic being carried over digital networks, the receiver threshold was considered to be 10^{-3} because below this, speech was unintelligible. In cable systems, the threshold is a semi-permanent error rate, often caused by cross-talk between cable pairs. When data became more mainstream, this 10^{-3} cutoff threshold was considered insufficient, and so 10^{-6} became the accepted threshold cutoff. ATM required a background error

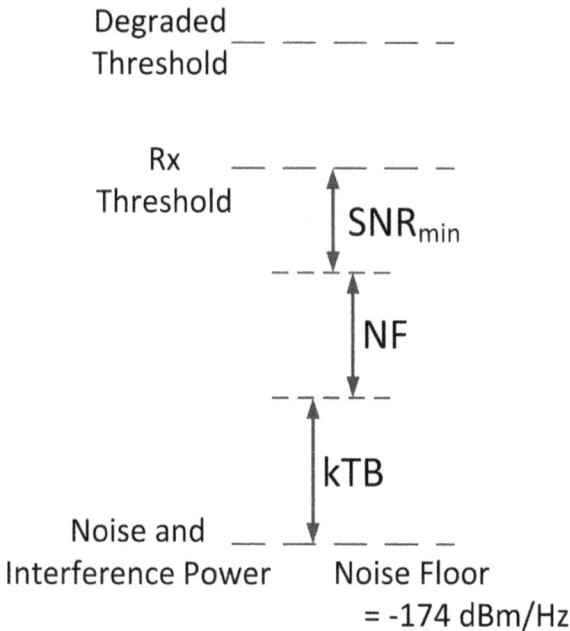

Figure 8.5 Receiver threshold composition

rate of better than 10^{-9}, although this was never standardised. The reason was that ATM usually ran over fibre, which achieved better than 10^{-11} and so performance was not an issue. The performance of fibre is so good that fibre systems are often considered error-free even though technically speaking an error rate of 10^{-13} still means that there is an error every 10^{13} bits. For the practical purposes of this book, we will consider this exceptionally good background error rate, an error-free region.

Microwave radio manufacturers realised they were competing with fibre optic systems, and so developed Forward Error Correction (FEC) systems that achieved a similar background error rate to fibre, making the 10^{-3} or 10^{-6} threshold cutoff virtually irrelevant.

Unfortunately, this is where the theoretical and practical world's clash. Referring to Figure 8.4, the so-called knee of the receiver threshold curve is the area between the horizontal background error-rate curve and the nearly vertical outage curve. When the radio receive level gets close to the threshold cutoff point, the error rate gets worse, and the FEC can no longer cope, and so additional errors are introduced which worries operators who compare radio and fibre systems. Arguments ensue about whether the 10^{-3} or 10^{-6} threshold should be used. In fact, when ATM first came out, concerns arose in the industry about whether 10^{-9} should be introduced as the threshold.

For most modern microwave systems the difference between the background error rate (usually 10^{-13}) and the 10^{-3} outage level is around 2 dB, making the cut-off issue completely irrelevant from a practical point of view. In fibre, the connection is either error-free or cut. In radio systems, for all practical purposes, it is the same.

In practical terms, the radio runs practically error-free, or is in a 10^{-3} outage, just like fibre, for all but this 2 dB window just before cutoff. The only practical benefit of the 10^{-6} threshold is for testing purposes in the laboratory. It has no practical significance at all in operational links, as the radio is incapable of discerning between a 10^{-6} or 10^{-3} fade given the speed that the fade moves past these markers.

For any radio planners reading this, I would still design the link using the 10^{-6} parameters as most people do, because the above argument is too challenging to explain to non-microwave practitioners, and the error introduced by using the *wrong* figure is not worth the pain of the debate.

LINK BUDGET

As discussed earlier, the fade margin is derived from the difference between the receiver threshold and the nominal receive level. The nominal receive level is determined through adding up all the losses and gains from the transmitter at one end, through to the receiver at the other end. This process is called the link budget.

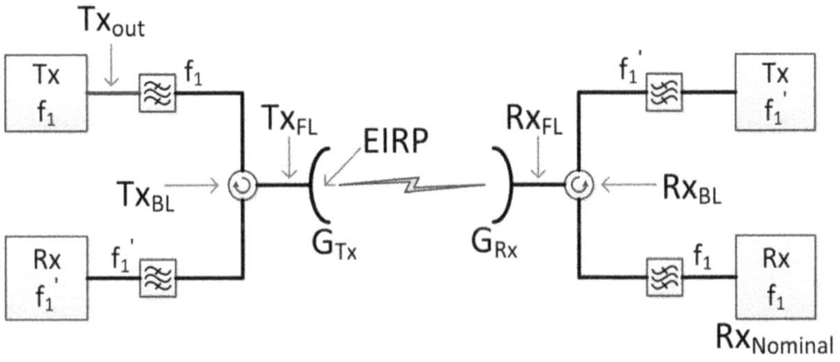

Figure 8.6 Link budget

The link budget elements are illustrated in Figure 8.6.

$$Rx_{Nominal} \text{ [dBm]} = EIRP - FSL \text{ (path)} - Aa + G_{Rx} - Rx_{BL} - Rx_{FL}$$

where

EIRP [dBm] = Effective Isotropic Radiated Power
$\qquad\qquad$ = Output power at the transmit antenna
$\qquad\qquad$ = $Tx_{out} - Tx_{BL} - Tx_{FL} + G_{Tx}$
$\qquad\qquad$ Tx_{BL} is Transmit Branching Loss
$\qquad\qquad$ Tx_{FL} is Transmit Feeder Loss
$\qquad\qquad$ G_{Tx} is Tx antenna gain
$\qquad$ FSL = Free Space Loss
$\qquad\;$ Aa = atmospheric absorption loss
$\qquad$ G_{Rx} = Rx Antenna Gain
$\qquad\qquad$ Rx_{BL} is Receive Branching Loss
$\qquad\qquad$ Rx_{FL} is Receive Feeder Loss

Free Space Loss (FSL) is the loss over the path between the two antennas unaffected by the earth, or any atmospheric and weather impairments. Signal strength falls off by the square of the distance between the two antennas. In the FSL equation, frequency gets introduced because the equation is derived from two isotropic antennas, and therefore wavelength (and hence frequency) becomes a factor.

$$FSL \text{ [dBm]} = 92.4 \text{ dB} + 20 \log (f \text{ GHz}) + 20 \log d \text{ (km)}$$

ANTENNA HEIGHTS

In chapter 4, we discussed diffraction loss. The extent of diffraction loss is a function of the clearance over the obstruction, as well as the type of obstacle causing the blocking. The design approach is to set the antennas at a height that diffraction is acceptable under all expected k-factor values. When the antennas are placed so that clearance over the dominant obstacle is set at precisely 60 per cent of the first Fresnel zone, diffraction loss is zero. Due to practical considerations as well as the fact that k-factor changes, this is not always possible, but the ITU (ITU-R P.530) has published a set of clearance rules to assist radio planners in achieving risk-free antenna heights as shown in Table 8.2.

Table 8.2

Clearance rules specified by the ITU

Rule 1	**Rule 2:** *Temperate climate*	**Rule 2:** *Tropical climate*
100% F1	0% F1 for isolated obstacle 30% F1 for extended obstacle	60% F1
K median (4/3)	Kmin (use hop length curve)	Not specified (⅔ is default)

For the diversity antenna, some diffraction loss can be accommodated even under median k conditions. By allowing some diffraction, the upper antenna can be kept reasonably low while still maintaining the 10-15 m spacing that is usually required for effective space diversity protection against multipath. It is still important to ensure that near-field obstacles are cleared by the diversity antenna. The lower diversity antenna positioning is possible because the antenna is required during multipath conditions, which is when k is very high, not low. If diversity is used for reflection protection, it is also beneficial for the antennas to be lower down, as discussed later.

In practice, there are a few challenges with applying the ITU clearance rules, which are listed below, but this only really has a significant impact on links longer than about 30 km.

- Antenna heights cannot be changed as the k-factor changes; therefore, when antenna heights are chosen at a height to optimise diffraction loss, usually non-optimal performance against multipath occurs, and vice versa.
- Diffraction analysis requires very precise path data which is only achievable if accurate profile height measurements are taken, which is time-consuming and costly to do.
- Protection against minimum k can result in the antenna being set at the very worst clearance condition (100 per cent F2) under median k conditions. This is especially true when following the rules for a tropical climate.

Given the considerations above, providing the radio designer has sufficient expertise, it may be preferable to set the antenna at the best height for nominal conditions and allow some diffraction loss under low k conditions. In practice, antennas often have to be set at a height based on practical issues such as antenna bracket availability, maintenance and installation constraints, commercial factors, safety concerns and site sharing constraints. Unless a *new* structure is being built, the office-based calculations are thus often used to validate options on a tower, rather than specifying the exact position to be installed. It is hoped that in this book enough theory has been covered to prove that antenna height placement is not an exact science and that the best height on the tower is the one that delivers the best performance overall, taking *all* considerations into account.

MULTIPATH OUTAGE

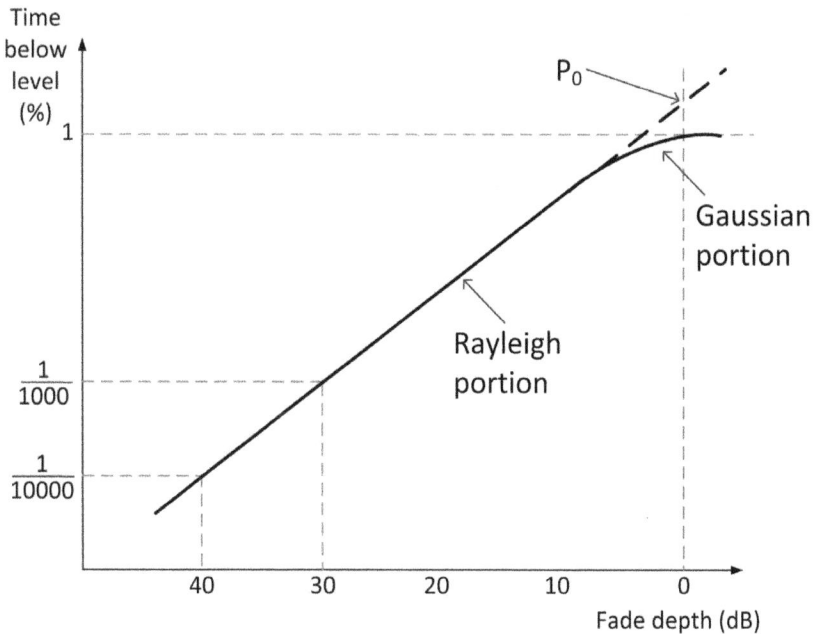

Figure 8.7 Rayleigh fading curve

The basic principle of estimating the so-called flat component of multipath fading is based on the assumption that fading follows a Rayleigh distribution curve, which is shown in Figure 8.7.

What this shows is that for every decade drop in RSL from fading, there is a tenfold probability decrease of that drop occurring. In other words, a 20 dB fade is ten times less likely to occur than a 10 dB fade. On a particular hop, providing you know the y-intercept point of the straight line, the so-called Po value, you can calculate the associated expected outage for any given fade margin, as shown in Figure 8.7. The Po value represents the inherent amount of fading expected and is called the *multipath fading occurrence factor*. Over the years, various empirical formulas have been developed to try to predict the Po value, as it has the most significant impact on multipath fading prediction accuracy.

According to ITU-R 530, revision 16, the multipath outage from flat fading is a function of Pw (also called Po) using a formula that includes the variables below:

$$Pw\,[\%] = K \cdot d^{3.4} \cdot (1+|\varepsilon p|)^{-1.03} \cdot f^{0.8} \times 10^{-0.00076\,hL - A/10}$$

$$K = \text{geoclimatic factor} = 10^{-4.6 - 0.0027 dN1}\,(10 + s_a)^{-0.46}$$

where

dN1 = point gradient of refractivity (G) not exceeded for 1% of average year

Values of dN1 can be obtained from ITU-R P.453

s_a = roughness factor (standard deviation of terrain heights in m)

εp = path inclination

d = distance (approximately cubed)

f = frequency

h_L = altitude of lower antenna

A = Flat Fade Margin (dB)

An alternative and much simpler model is used in the US, based on the Barnett-Vigants formula.

$$Po = A . f . d^3 . 10^{-3}/ S_1^{1.3}$$

where

A = climatic factor between 1 (mountains) and 4 (coast)
f = frequency
d = distance
S1 = roughness factor

RAIN OUTAGE

On short hops, the main performance variable is rain fading. The Receive Signal Level (RSL) of the radio is reduced due to absorption and scattering. The effect gets worse the smaller the wavelength is, and hence attenuation increases with frequency. In some areas, wet snow can also impact links, similar to rain. In tropical regions, rain even needs to be considered below 10 GHz, but usually, rain has its most significant impact for bands above 10 GHz. In practice, rain impacts the usable hop length, and it is the instantaneous rainfall rate, rather than annual rainfall, that is of interest. The design methodology takes into account these rain rates, which are published as rain regions, to predict the expected outages. The basic approach is to find the rain attenuation from the rain rate (mm/hr) required to fade the path, and then determine how often this rain rate occurs for that geographic area. Radio planning software is usually required to calculate the exact values using the formulas from ITU-R P.530.

In ITU-R P.530, the formulas are provided to calculate the expected outage, and rain maps are provided as part of that methodology. The prediction methodology is relatively complex. Rain cells get smaller the higher the intensity of the rain. Hence the attenuation from water absorption cannot be evenly distributed over the entire physical path length. An effective path length needs to be determined over which absorption is calculated.

In the US and Canada, the Crane methods and rain maps are often more comprehensive than the ITU.

Once again, practical considerations often determine the antenna size to be used, and the formulas in the radio software are used to validate the expected performance and outage risks. Where radio planning predictions are used to specify the requirements, without taking into account practical considerations, and limitations, and tolerances of the prediction formulas, unnecessary costs can be incurred without the benefit of increased performance.

COUNTERMEASURES FOR LINK FADING

As already mentioned, microwave radio links can be designed to be extremely stable and achieve carrier-grade availability. By designing the system with adequate fade margin, the link will perform well, even in poor weather. To overcome so-called flat fading, where the RSL falls below the nominal level, antennas are sized accordingly to meet a specified fade margin. Adequate system gain protects the system against rain fading as well as the flat component of multipath fading. Good adaptive equalisation in the radio equipment overcomes the selective element of multipath fading. If even the largest antenna sizes are inadequate for rain fading, a lower frequency band will decrease the outage time.

For very long hops, typically over 50 km, space diversity can be used to overcome the signal reduction effects of multipath fading. Space diversity with a very specific antenna spacing is used against specular reflections.

Choosing apparatus licences, and high-performance antennas will ensure that interference is minimised. It is also good practice to put the antennas as low as diffraction losses will allow, to reduce the amount of interference experienced and generated.

OVERALL RELIABILITY

Customers care about their service performance, far more than they do about percentage claims of network reliability. Claims of 99.99x per cent may look good on a brochure, but they are meaningless if the telecoms link to a major hospital is down, thus putting someone's life at risk! No transmission medium on its own can achieve 100 per cent availability. It has already been mentioned that even fibre optic cables, which are the gold standard, experience physical breaks, and when it happens, the outage can be extensive. The utopia situation is to have virtually error-free performance when the service is available and when a cut occurs, for the service to automatically continue on a redundant link. Carrier-grade microwave radio provides the same, virtually errorless, performance as fibre if designed and installed correctly. Fibre optics and microwave radio thus make an ideal combination, as they do not share any common infrastructure, thus ensuring that any break in service on either link, will not occur on the backup link.

A microwave hop design should be considered holistically in terms of how, and where, it is being deployed in the network. The hop reliability figures produced by most radio planning software tools tend to focus only on the theoretical propagation outages for the individual hop itself and ignores site outages, including local power supplies, site outages, as well as multi-hop correlated events. It should be remembered that in the original ITU radio link standards, targets were set for a section of radio hops and not assumed to be applied to an individual hop. As already discussed, unfortunately there are currently no easily-applied multi-hop standards that can be used, but planners should be careful not to just add up individual hop calculations from a software planning tool and aggregate them and then apply this to a multi-hop link of radio hops. Where outages are correlated between hops, this should be taken into account in the overall design.

CHAPTER IN A NUTSHELL AND FINAL CONCLUSIONS

Establishing a microwave radio link and getting it to work error free is incredibly easy. So long as the path is not completely blocked and you point the antenna roughly in the right direction, not only will the link be established, but it will run error free. The challenge is that when atmospheric effects and weather changes stress the link, errors and unreliability may occur. To get a microwave link running error-free, even when it is pouring with rain or when abnormal layers of hot, humid air bend the radio signal, requires some expertise in planning and deployment.

For the millimetre wave bands, *obstruction losses* and *rain* are the main variables that impact performance. A physical line-of-sight check should be done and then the antennas placed at an appropriate height so that there are a few metres of unobstructed clearance around the centre point of the beam across the entire path. All the theory about k-factor and refraction and diffraction losses are not relevant to most paths in these higher frequency bands, as the link lengths are too short for these additional variables to have much practical significance. Picking the right rain rate or region, to predict how often the rain intensity will fade the path down to the receiver threshold, is the main consideration in achieving high availability. In addition, interference should be considered, including high-low bucking planning to ensure that the receiver threshold is not degraded.

On longer links, and especially for links exceeding 30 km, the process of predicting performance due to multipath fading is much more complex, and all the analysis of k-factors and Fresnel zones is very relevant.

If you get it right, you can design an ultra-reliable link operating over hundreds of kilometres. This statement is not an exaggeration. In Australia, the author, together with his colleagues in a private network operator - specialising in building networks for critical applications such as those found in health, education and public safety - built a high capacity link exceeding 200 km, which carries live traffic and exceeds five-nines network availability. It should also be pointed out that this link was established in a tropical area - the worst area for fading - with small dishes placed quite low down on the tower.

In terms of so-called five nines carrier grade availability, the most important thing is to have some form of diversity both in terms of protecting hardware failure as well as path protection. Ultra-high availability can be achieved by ensuring that the multiple telecoms paths have no common points of failure. Using microwave as a backup to fibre, or other telecoms transmission media, creates an ideal ultra-reliable network connection.

With the advent of 5G and Internet of Things (IoT), as well as the myriad of current applications, microwave radio is guaranteed to be around for a long time. It is hoped that the information provided in this booklet will be a helpful, quick reference guide for those who design, operate and use telecoms services, running over microwave radio.

LIST OF ACRONYMS AND ABBREVIATIONS

5G	5th Generation cellular networks
ACM	Adaptive Coding and Modulation
ADM	Add-Drop Multiplexer
ADSL	Asymmetric Digital Subscriber Loop
API	Application Programming Interface
ATM	Asynchronous Transfer Mode
ATPC	Automatic Transmit Power Control
BCA	Band and Carrier Aggregation
BPSK	Binary Phase Shift Keying
BER	Bit Error Rate
BBER	Background Bit Error Rate
CCDP	Co-Channel Dual Polarisation
C/I	Carrier to Interference ratio
dB	decibels
DEM	Digital Elevation Model
DFM	Dispersive Fade Margin
DL	Diffraction Loss
DOCSIS	Data Over Cable Service Interface Specification
DSL	Digital Subscriber Loop
DSM	Digital Surface Model

DTM	Digital Terrain Model
EFM	Effective Fade Margin
EHF	Extremely High Frequency
EIRP	Effective Isotropic Radiated Power
EM	ElectroMagnetic
ETSI	European Telecommunications Standards Institute
F/B	Front-to-Back ratio
FD	Frequency Diversity
FDD	Frequency Division Duplexing
FFM	Flat Fade Margin
FM	Frequency Modulated (also sometimes used for Fade Margin)
FSK	Frequency Shift Keying
FSO	Free Space Optics
FSL	Free Space Loss
GDA	Geocentric Datum of Australia
GE	Gig Ethernet
GHz	GigaHertz (10^9)
GIS	Geographical Information System
GPS	Global Positioning System
H-QoS	Hierarchical Quality of Service
HP	High Performance
HPBW	Half Power BeamWidth
HSB	Hot StandBy
IAB	Integrated Access and Backhaul
ICNIRP	International Commission on Non-Ionizing Radiation Protection
IDU	InDoor Unit
IEEE	Institute of Electrical and Electronic Engineers
IF	Intermediate Frequency

IFM	Interference Fade Margin
IMP	InterModulation Products
IoT	Internet of Things
IP	Internet Protocol
IP-MPLS	Internet Protocol MultiProtocol Label Switching
ISI	InterSymbol Interference
ITU	International Telecommunication Union
ITU-D	ITU Development agency
ITU-R	ITU Radiocommunications agency
ITU-T	ITU Telecommunications agency
LAN	Local Area Network
LEO	Low Earth Orbit
LiDAR	Light Detection and Ranging
LOS	Line of Sight
LSO	Lifecycle Service Orchestration
MAC	Media Access Control
Mbps	Megabits per second
MEF	Metro Ethernet Forum
MEO	Medium Earth Orbit
MHz	MegaHertz (10^6)
MIMO	Multiple Input Multiple Output
MODEM	MOdulator-DEModulator
MPLS	MultiProtocol Label Switching
MTBF	Mean Time Between Failure
MTTR	Mean Time To Repair
MULDEM	MULtiplexer-DEMultiplexer
NFD	Net Filter Discrimination
NFV	Network Function Virtualisation
NIR	Non-Ionising Radiation
N-LOS	Non Line-Of-Sight
nm	nanometre (10^{-9})

NP	Non Protected
NTP	Network Time Protocol
ODU	OutDoor Unit
OMT	OrthoMode Transducer
PEP	Path End Point
PDH	Plesiochronous Digital Hierarchy
PMP	Point to MultiPoint
POP	Point Of Presence
PSK	Phase Shift Keying
PTP	Point To Point
QAM	Quadrature Amplitude Modulation
QPSK	Quadrature Phase Shift Keying
QoS	Quality of Service
RF	Radio Frequency
ROI	Return On Investment
RPE	Radiation Pattern Envelope
RSL	Receive Signal Level
RTPC	Remote Transmit Power Control
RX	Receiver
SDH	Synchronous Digital Hierarchy
SDSL	Symmetric Digital Subscriber Loop
SDN	Software Defined Networks
SD-WAN	Software Defined Wide Area Network
SES	Severely Errored Seconds
SHF	Super High Frequency
SLA	Service Level Agreement
SNR	Signal to Noise Ratio
SONET	Synchronous Optical NETworks
SRTM	Shuttle Radar Topography Mission
SWR	Standing Wave Ratio
TAE	Transverse Adaptive Equaliser

TCO	Total Cost of Ownership
TDD	Time Division Duplexing
TDM	Time Division Multiplexing
TEM	Transverse ElectroMagnetic
TX	Transmitter
UHF	Ultra-High Frequency
UI	Unit Intervals
VDSL	Very high bit rate Digital Subscriber Loop
VSWR	Voltage Standing Wave Ratio
WAN	Wide Area Network
WGS	World Geodetic System
Wi-Fi	Wireless Fidelity
WRC	World Radio Congress
XPD	Cross-Polar Discrimination
XPIC	Cross-Polar Interference Canceller

APPENDIX USEFUL FORMULAE

Wavelength

$\lambda = c/f$

$\lambda = 300/f$ (MHz) m

$\lambda = 1/f$ (GHz) ft

Free Space Loss (FSL)

FSL = 36.6 + 20 log d (miles) + 20 log f (MHz) dB

FSL = 96.6 + 20 log d (miles) + 20 log f (GHz) dB

FSL = 92.4 + 20 log d (km) + 20 log f (GHz) dB

First Fresnel Zone

$F_1 = \sqrt{}$ (λ (d1.d2) / (d1+d2))

$F_1 = 17.3\sqrt{}$((1/(f (GHz)).((d1.d2)/(d1+d2))) m Note: (d1 and d2 in km)

F_1 (midpoint) = 17.3 $\sqrt{}$(d1/2f (GHz)) m

$F_n = F_1 \sqrt{n}$

Parabolic Antenna Gain (G)

G dBi = 18 + 20 . (log (D (m).f(GHz))

G dBi = -43 + 20 . (log (D (m).f(MHz))

G dBi = 7.35 + 20 . (log (D (ft).f(GHz))

Half Power Beamwidth (HPBW)

$$\text{HPBW} = 163 / \sqrt{10}^{\,A\text{dB}/10}$$

Near-Field (NF)

$$\text{NF} = 6.7 \cdot D^2 (m) \cdot f_{GHz}$$

Refractivity Gradient and k

$$k = (157/(157 + G))$$

Velocity Factor

$$\text{VF} = 1/\sqrt{\varepsilon}$$

Fading formulae (Po)

ITU: $\text{Po } [\%] = K \cdot d^{3.4} \cdot (1+ |\varepsilon p|)^{-1.03} \cdot f^{0.8} \times 10^{\,-0.00076\, hL - A/10}$

where $K = 10^{\,-4.6-0.0027 dN1} (10+s_a)^{-0.46}$

US: $\text{Po} = A \cdot f \cdot d^3 \cdot 10^{-3}/ S_1^{\,1.3}$

USEFUL REFERENCES

BOOKS

Crane, R 2003, *Propagation Handbook for Wireless Communication System Design,* CRC Press.

Freeman, R 1987, *Radio System Design for Telecommunications,* John Wiley and sons.

Henne, I & Thorvaldsen P 1999, *Planning of line-of-sight radio relay systems*, Nera.

Huurdeman, A 1995, *Radio-Relay Systems,* Artech House.

International Telecommunication Union, 1996, *Handbook Digital Radio-Relay Systems,* ITU Radiocommunication Bureau.

Manning, T 2009, *Microwave Radio Transmission Design Guide*, 2nd edn, Artech House.

Wells, J 2010, *Multi-Gigabit Microwave and Millimeter-Wave Wireless Communications,* Artech House.

PAPERS AND ARTICLES

Edstam, J 2016, 'Microwave backhaul gets a boost with multiband', *Ericsson Technology Review.*

ETSI, February 2018, 'Microwave and Millimetre - wave for 5G Transport', *ETSI White paper No.25.*

ETSI GR, November 2017, 'Frequency Bands and Carrier Aggregation Systems', *mWT 015.*

FCC, March 2019, 'Title 47. Electronic Code of Federal Regulations', *www.fcc.gov* (Note: Paper refers to General licensing rules)

GSMA, April 2018, 'Road to 5G: Introduction and migration', *www.gsma.com*

Hilt, Attila et al. 2008, 'Access transmission network upgrade in a nationwide mobile network modernization project for EDGE deployment', *The 13th International Telecommunications Network Strategy and Planning Symposium.*

Hilt, A & Petras, P May 2007, 'Microwave Transmission Node Optimisation for Access Capacity Increase in Mobile Networks', *Proceedings of Microwave Optical Week.*

Manning, T & Little, S November 2009, 'Deployment and link planning of Adaptive Coding and Modulation radio networks', *Microwave Journal.*

Manning, T August 2002, 'Microwave Radio Link Performance', *Applied Microwave and Wireless.*

MEF white paper, January 2011, 'Microwave Technologies for Carrier Ethernet Services', *www.mef.net*

OFCOM, July 2010, 'Short range devices information sheet', *www.ofcom.org.uk* (Note: Paper refers to License exempt devices)

SUPPLIER WHITE PAPERS

Aviat, August 2019, *'Capacity + reliability multi-band is the right solution for 5G backhaul'.*

Ceragon, 2016, *'Advanced Frequency Reuse - More Capacity Out of Current Spectrum'.*

Ericsson, December 2018, *'Ericsson Microwave Outlook'.*

Ericsson, Ericsson Technology Review 2017, *'Microwave Backhaul beyond 100 GHz'.*

NEC, February 2018, 'Evolving microwave mobile backhaul for next-generation networks'.

Nokia, 2019, 'The evolution of microwave transport - enabling 5G and beyond'.

STANDARDS

Standard	Topic of standard
IEEE 802.3	Ethernet standard
IEEE 1588.v2	Packet synchronisation
ITU-R F.452	Interference
ITU-R F.746	Frequency arrangements
ITU.R F.2113	Performance and Availability
ITU-R Report F.2323	Future trends
ITU-R Report F.2416	Future frequency band (275 - 450 GHz)
ITU-R P.310	Propagation definitions
ITU-R P.452	Interference prediction
ITU-R P.453	Refractivity data
ITU-R P.525	Free space attenuation
ITU-R P.526	Diffraction

ITU-R P.530	Main microwave radio design
ITU-R P.676	Specific attenuation data
ITU-R P.835	Standard atmosphere data
ITU-R P.837	Rain modelling
ITU-R P.838	Specific attenuation for rain
ITU-T G.828	Quality and availability targets
ITU-T G.8260	Synchronisation definitions

ABOUT THE AUTHOR

Trevor Manning has a long history of working in the Microwave radio field. Following his graduation as an Electronics engineer with a specialisation in Microwave, he spent some time in Italy working in the microwave radio research department of what is now Nokia. Working closely with the chairman of the ITU working group that wrote ITU-R P.530 - the main design document for radio link design - he gained both practical and theoretical insights into radio link design. These insights allowed him to design a very innovative microwave network in South Africa as part of a hundred million dollar network upgrade, which has stood the test of time. He went on to have national responsibility for the maintenance and operation of the network before moving to the UK to work for an American microwave manufacturer.

Trevor currently lives in Brisbane, Australia where his focus is on Leadership training of technical people but is still actively involved in innovative Microwave design. He recently led a team of radio engineers to design and build the world's longest high capacity Ethernet link. Spanning over 200 km, this link achieves 99.9x% before diversity improvements, despite being located in a tropical part of Australia. The real innovation is the fact that it has very small antennas installed low down on the tower, breaking all the conventional rules of radio link design.

ACKNOWLEDGEMENTS

When writing a book there are so many things that you think you have written, or concepts that you believe you have clearly explained, that often do not *actually* appear in the words you have used in the book.

Feedback from beta readers provided insights into what I have written, compared to what I intended to write, as well as providing additional information and expertise, tapping into broad experiences on an international basis.

Many people provided useful input to this book, with no promise of any personal rewards, so it is hard to know where to start expressing my gratitude.

I want to thank Ian Sutton from Ceragon who, in addition to providing invaluable industry commentary, has inspired me to write a book from the heart, rather than just a boring engineering textbook. He encouraged me to break from tradition and include personal stories and insights. I made a conscious decision to write these stories in the first person, instead of following academic grammatical pedantic-ness. I hope that this has created a more enjoyable and easy-to-read book.

I also want to thank Professor Andy Sutton from BT who provided useful feedback on the first draft of the book and put me straight on some 5G concepts that I had attempted to include. I have been critical of the many myths and inaccuracies in mainstream microwave material out there, and I was falling down the same trap with 5G. Rather than regurgitating what others have written on 5G, some of it erroneous, I have deliberately stuck to my area of expertise and will leave it to someone else to write a book on 5G.

Dr Attila Hilt from Nokia has a fantastic eye for detail and provided me with pages and pages of useful feedback and corrections from my earlier drafts. Dr Jonas Edstam from Ericsson has also been incredibly helpful in sharing his expertise and experience on a range of topics covered in the book. Likewise, Oliver Bosshard (Real Wireless), Steve Odell (BT) and Marcel Vonarburg (Senior industry consultant) provided detailed feedback that has positively contributed to the technical accuracy of the book.

I would like to thank everyone else who contributed as beta and advanced readers of the book. There are too many to name, but in particular I would like to mention Stuart Little (Aviat), Krisztian Som (NEC), Gunnar Nilson (NEC), Eddie Stephanou (Cambium), Steve Jones (Orange), Steve Simpson (KTL), Bryce Fisher (Telent), JJ Blignaut (Ericsson), Lloyd Mphahlele (MTN), Vitalijs Lopatinskis (SAF Tehnika) and Edgars Dudko (SAF Tehnika). Thanks also to Mark Arkell from the Australian frequency regulator, who provided useful insights on the terminology used internationally for licensing.

Sylvester Singarayer provided very useful and practical feedback on current equipment and field realities, as well as doing an outstanding job of creating professional looking diagrams for the book, from my rough pencil sketches.

Leaving the best for last, I will be eternally grateful to my wife Berry for the countless hours she has spent listening to my ideas, challenging my grammar, and handling the editing elements of the book from the cradle to the grave.

I feel like the formula one driver who might drive solo, but who couldn't complete the race without a great team behind them. To everyone who has contributed to creating what is hopefully a helpful and up-to-date industry reference book, I am grateful.

ENDORSEMENTS

'An easy-read for all new planners, but also a great reference handbook for experienced planners. A good balance of theory and real-world, practical considerations'. Oliver Bosshard, COO, Real Wireless

'A lot has happened to microwave over the years, with a lot still emerging. This short book has managed to provide a comprehensive overview of both old and new, which is not an easy task'. Dr. Jonas Edstam, Portfolio Strategy Manager, Ericsson

'This is a great addition to Trevor's more in-depth design guide, which is still a constant companion. This was well written, well thought out, and the personal touch makes this a surprisingly good read for a relatively academic subject'. Bryce Fisher, Wireless Technical Design Authority, Telent

'I would not often describe a book on microwave radio systems as a "page turner", but I found I wanted to keep reading'. Adrian Grilli, Associate, European Utility Telecoms Council

'*Several scientific papers and microwave books publish deep theory with long equations, leaving design engineers lost or confused when planning practical networks. This book helps designers gain an essential understanding of the background and physics behind the empirical formulas programmed into microwave software, so that designs can be adapted to local conditions.*' Dr Attila Hilt, Senior Systems Specialist, Nokia

'*A practical and concise approach to a complex topic, comprehensively dealing with the main issues and pitfalls to be encountered in Microwave link planning. An ideal ready reference for all involved in MW design, planning and deployment*'. Steve Jones, Senior Architect, Wireless Transport Networks, Orange

'*Extremely effective "how-to" for microwave link design. An expert mix of practical experience and background theory. Good for the new engineer or the experienced professional*'. Paul Kennard, Senior Consultant, McKay Brother International

'*The book is so well written that reading it feels like getting advice from some "microwave mate" on this complicated and at times abstract phenomenon. It describes the most important concepts and principles in the microwave world in simple and friendly language and is intertwined with the author's personal experiences around the globe, which keeps the reader's interest.*' Vitalijs Lopatinskis, Director of Technical support team, SAF Tehnika

'*This book truly covers the A-Z of microwave planning. The advice, if followed, would make microwave networks a lot more reliable and trusted than what they are today'.* Lloyd Mphahlele, General manager transport, MTN

'At last, a handy and affordable reference guide which demystifies the technicalities of designing and planning microwave and millimetre wave radio links, this is a must have book for anyone with an interest in wireless transmission and mobile backhaul'. Professor Andy Sutton, Principal Network Architect, BT

'If I look back at all my years working in telecoms, the most notable advice and information which remains foremost in my mind, are those based on practical experience and field observation. This book taps into Trevor's decades of personal & real-world experiences and so can be a great reference source for engineers in the field of wireless communication technology'. Ian Sutton, Technical Director, Ceragon.

'An excellent reference book for microwave planners. Without going into too much detail it addresses all the aspects that one has to consider when working for an Operator, as a planner, or as a microwave installer'. Pierre Redelinghuys, Senior Consultant, Technology Enterprise planning, MTN

'A fun and easy read that brings across the most pertinent points regarding planning microwave links'. Marcel Vonarburg, Senior Consultant

OTHER BOOKS BY TREVOR MANNING

Microwave Radio Transmission Design Guide - *Publisher: Artech House*

Help! What's the secret to leading engineers - *Order from Amazon*

Help! I need to master critical conversations - *Order from Amazon*

Help! They made me the team-leader - *Order from Amazon* (2020)

TRAINING COURSES AVAILABLE

Contact Trevor Manning to enquire about 1 day refresher training as well as an intensive 3 day masterclass in various locations around the world.

MICROWAVE DESIGN EXCEL CALCULATOR AVAILABLE

Email Trevor Manning to receive a free microwave design spreadsheet calculator covering most formulas used in the book. author@tmcglobal.com.au

Milton Keynes UK
Ingram Content Group UK Ltd.
UKHW010155100124
435758UK00003B/25